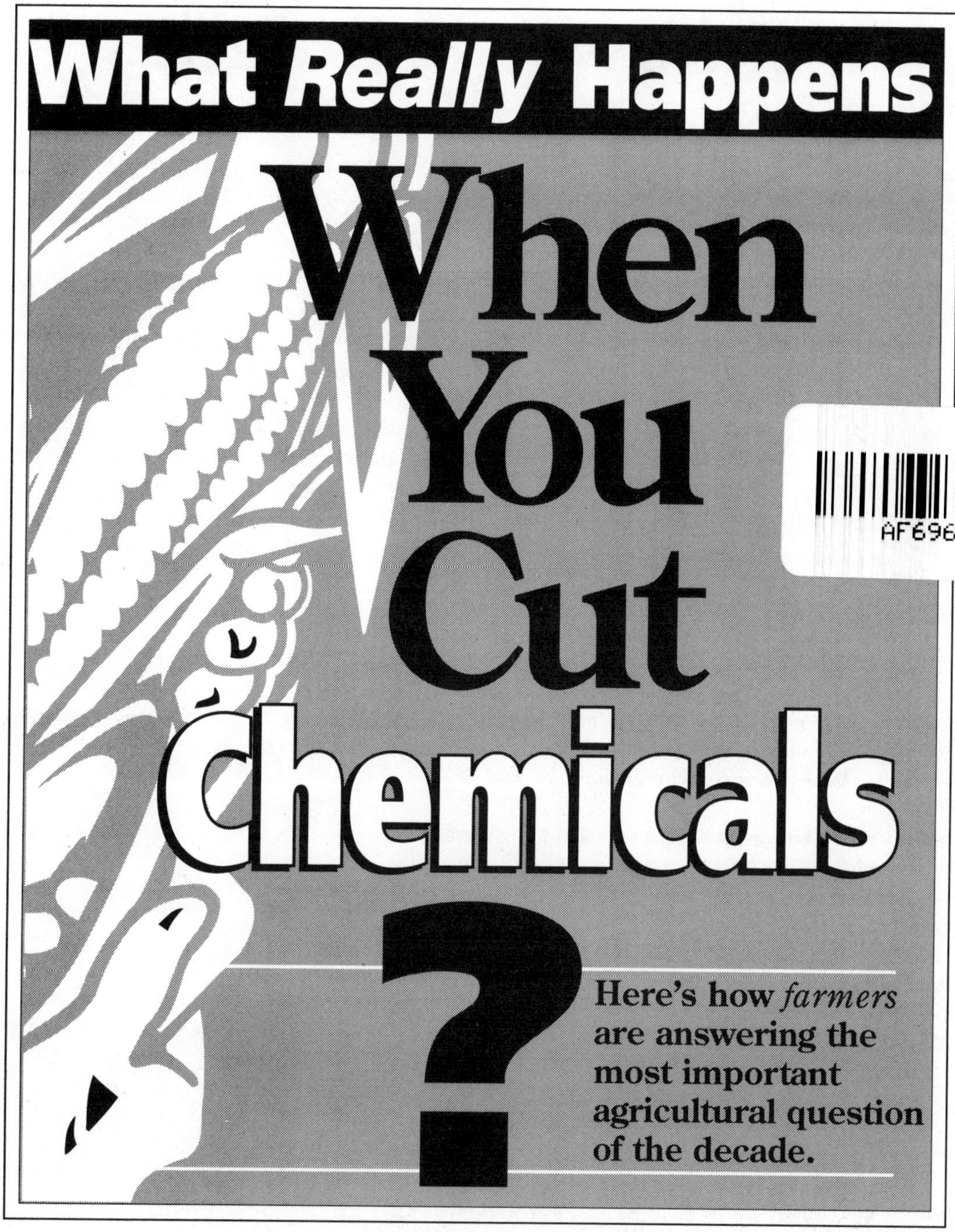

Edited by Christopher Shirley
and the staff of
The New Farm
Greg Bowman
Mike Brusko
Karen Cicero
Craig Cramer
T.L. Gettings
Bob Hofstetter

Acknowledgements

Special thanks are due the farmers and researchers featured in this book. They gave freely of their time talking with the editors and reviewing manuscripts. Thanks also are due the Rodale Institute national network coordinators (Dr. Jim Tjepkema, Leon Weber and Janet Bachmann); Tom Gettings and his photo staff at Rodale Press and Rodale Stock Images; the Rodale Institute development office team (especially Tina Abel, Jennifer Eddinger, Kris Siessmayer and Pat Moyer); Deb Protchko, Chris Klipple and the Rodale Press marketing team (Laura Herzog, Tracy Hambleton, Susan Kelly and Maria Rodale); Helen Clogston and Jodi Schaffer from Rodale Books; former *New Farm* editor George DeVault (now editor of the Russian version, *Novii Fermer*); former FONE information coordinator Anne Schauer; and countless others who provided additional photos, information and encouragement.

A grant from the Charles Stewart Mott Foundation helped make this book possible. Support from the Mott Foundation, Rockefeller Brothers Fund and individual donors helped establish Rodale Institute's Midwest Farmer Network. Additional funding for Rodale Institute's national networks has been provided by The Joyce Foundation, the Ruth Mott Fund, The Winthrop Rockefeller Foundation, Winrock International, the Kerr Center for Sustainable Agriculture, the Jessie Smith Noyes Foundation, the USDA-SARE program, state project grants, donor contributions and the sale of Rodale Institute books and periodicals.

We also recognize the impressive work being sponsored by the Northwest Area Foundation, the Wisconsin Rural Development Center, the Center for Rural Affairs in Walthill, Neb., the Land Stewardship Project in Lewiston, Minn., Practical Farmers of Iowa and other organizations that are helping farmers rediscover, and develop further, reduced-chemical farming techniques. As catalysts for change, these people, groups and programs make possible the hundreds of field days and workshops held annually around the country to promote healthier farming methods.

Photography: T.L. Gettings, Rodale Press, Rodale Stock Images
Text Design and Page Layout: Karen Cicero and Karen Bleiker
Cover Design: Scott Protchko
Photo Research: Ramona Neidig and Rose Reichl, Rodale Stock Images

Published in the United States of America by Rodale Institute, 222 Main St., Emmaus PA 18098.

Library of Congress Cataloging-in-Publication Data

What really happens when you cut chemicals? / edited by Christopher Shirley and the staff of The New Farm, Greg Bowman ... [et al.].
p. cm.
Includes index.
ISBN 0-913107-16-6
1. Sustainable agriculture. 2. Sustainable agriculture—United States. I. Shirley, Christopher.
S494.5.S86W48 1993 92-41471
630—dc20 CIP

Table of Contents

Foreword

You may not know it yet, but buying this book makes you part of a nationwide conspiracy.

It's a conspiracy to raise food prices, cut crop yields and cause famine and disease. It's a conspiracy to put farmers out of business and to undermine all of the progress that American agriculture has made in the past century.

Nary a month goes by that we at *The New Farm* don't come across at least one scientific study forecasting all of those woes (and then some) if farm chemicals are banned.

Such studies make us feel about the same way you must have felt after reading the first sentence on this page. Angry. Insulted. Frustrated. Misunderstood.

Granted, many of the bugaboo forecasts could come true. Simply banning all farm chemicals probably would put our health and our food supply — not to mention our nation's farmers — in great jeopardy.

Still, to us, the fact that these studies continue to be done, and continue to be publicized in trade ads and farm magazines, does more harm than good to American agriculture.

The message they send is that "Our system ain't broke, so why fix it?"

"Our food is the safest and cheapest in the world," the litany goes. "Pesticides don't show up in wells anymore today than they did decades ago. And if they do, the levels often are so minuscule that you'd have to drink barrels of the so-called 'tainted' water just to break out in hives.

"Besides, there's no such thing as 'zero risk.'"

There are two things wrong with this message.

First, it assumes that farmers are seeking alternatives to agrichemicals *only to solve a problem*. This allows the opponents of low-input farming to beg the question and say, "There *is* no problem."

The result is predictable: Farmers and other agricultural professionals become distracted from the real benefit of seeking alternatives to chemicals, which is to *discover new opportunities*. Opportunities to regenerate our natural resources and rural communities. And opportunities to make farming more profitable over the long term.

The second thing wrong with this message is it assumes that banning chemicals is the underlying agenda of any person or group seeking to help farmers adopt low-input methods.

That's like assuming you want to ban automobiles if you encourage people to explore alternative forms of transportation, such as bicycles or shoe leather. It's like assuming you want to ban TV if you encourage your kids to entertain themselves by ... (gasp!) ... reading a book once in awhile.

Why spend millions of dollars and countless months asking about the potential risks of banning all farm chemicals — when there's no chance that such a ban would ever take place?

A much more likely scenario is that farm chemical use will drop dramatically in the coming years, as farmers seek new opportunities to improve their health, natural resources and profit margins. This trend already has begun.

So, the question those studies *should* be asking is the one that appears on the front cover of this book:

"What *really* happens when you cut chemicals?"

The good news is that people are starting to ask that question. And in this book, you're going to hear the answers they're getting.

Some of the answers come from long-term trials comparing high- and low-input cropping systems side by side under carefully controlled conditions. Once rare, such studies now are well-established at research sites run by nonprofit groups like Rodale Institute, as well as by land-grant universities, USDA and EPA. They'll give you the hard-and-fast data you need to help decide what chemicals you might be able to reduce, and how and when to start.

More and more farmers also are discovering what really happens when they cut chemicals on their own farms. They're doing their own side-by-side trials, often with help from trained researchers. Surveys of these farmers paint a statistical picture of low-input farming's potential risks and benefits on a farm similar to yours.

Finally, and most importantly, case studies of individual low-input farmers add a personal touch to the statistics. They provide insights into the challenges, opportunities, successes and failures that you just can't find on a computer printout.

Taken together, these three sources of information will give you what we hope are practical, accurate answers to *your* questions about cutting farm chemical use.

Or maybe they'll just fuel another conspiracy theory.

Mike Brusko
Publisher, *The New Farm*
November 1992

Chapter 1: Farmers Show The Way

"Maybe nature itself is the best miracle of all."

— Robert Rodale

To regenerative thinkers everywhere.

Here's 'a proposal for profit' based on reduced-chemical farming systems.

"Agriculture is like a three-legged stool," Bob Rodale said in 1983. "One leg is production — both food and fiber. Another leg is made up of the earth-based resources — soil and water. People form the third leg.

"The production leg of our agricultural stool is now quite long, and the other legs are short and growing shorter. As a result, society can no longer sit comfortably on the stool.... But, with planning, you'll be able to sit on it for a long time, and so can your children."

Bob's seating plan called for preserving and renewing the dwindling supply of what have long been American strengths: fertile fields and family farms. "Regeneration of the soil is needed now," he said in *The New Farm* magazine ("A Proposal For Profit," March/April 1985). "Mere conservation is no longer enough. Conservation and agriculture will have to merge into regenerative production systems." In this book, the most comprehensive of its kind, we'll introduce you to some of the farmers and researchers who are ushering in that new era of regenerative agriculture.

We'll start by looking at nine farms in the Rodale Institute national networks. Readers of *The New Farm* magazine will recognize some of the "farm heroes" here, as we've featured a number of them in past issues. They include cash-grain farmers, dairy and livestock producers and diversified growers whose operations run from a couple hundred acres to several thousand. In succeeding chapters, additional case studies and farmer comments help answer many questions you may have about reducing agrichemical use.

Hearing from experienced low-input farmers such as these — and noting any snags they overcame — makes it easier for you to test and fine-tune reduced-chemical methods on your farm. "The innovative power of individual farmers is tremendous and is doing more to solve basic production and other problems than any of the experts give it credit for," Bob Rodale realized years ago. "Largely on their own, farmers are cutting production costs, improving product quality and lessening their impact on the environment."

Bob recognized the intrinsic strengths of individuals *and* communities. "Just as farmland can improve using internal resources, so can we. All of us have assets and sources of inner strength that can be used more effectively," he wrote. Many of these talents can best be tapped and developed "only when linked with the missing pieces of resources in others."

He wasn't speaking wishfully. Over the years, Bob toured hundreds of successful farms where few off-farm inputs were used. Starting in the early '70s, he sponsored low-input farming research at a 305-acre experimental farm in southeastern Pennsylvania. To help farmers reduce their dependence on chemicals, he founded *The New Farm* magazine in '79. With the help of John Haberern and others at the nonprofit Rodale Institute, he encouraged Congress and the USDA to establish the Low-Input Sustainable Agriculture (LISA) program, now known as the Sustainable Agriculture Research and Education (SARE) program. He also helped Rodale Institute staff establish innovative on-farm research networks in the '80s.

For Farmers, With Farmers

When we started the national networks, many university researchers were looking down their noses at on-farm research. "You can't control variables sufficiently or generate valid numbers from on-farm data," they said.

We wanted to change that view. Farmers' informational needs often differ from university research priorities. On-farm tests using field-size strips to vary a single input — and broader studies of whole farms and farming systems — could provide practical information, we knew. And on-farm wisdom provides the building blocks needed to help other farmers reduce their dependence on purchased inputs.

That became more obvious to many farmers in the mid-'80s, when grain prices declined. The "green revolution" of the '60s and '70s carried a high price in the '80s: Even ignoring health risks associated with excessive use of agrichemicals, the cost of production began outpacing crop returns on many conventional farms. It was time, we knew, for farmers, researchers and communities to work together to regenerate America's farming systems.

Rodale Institute's Midwest Farmer Network started in '84, when we began sponsoring research conducted by Boone, Iowa, farmers Dick and Sharon Thompson. (For more about the Thompsons, see "The Last Chapter Hasn't Been Written," later in this book.)

The Midwest Network soon included a dozen farms, representing many different soil types and climatic conditions. Much of the research conducted on these farms focuses on legume cover crops to improve soil tilth and to supply nitrogen. Rotational grazing and pasture productivity are being evaluated, as well as the use of high-value crops to diversify and reduce risk.

We established a Mid-South Network in '87 to address production constraints related to warm climate, high rainfall and humidity, low organic-matter levels and heavy pest- and disease pressure. Reduced-chemical farming studies there involve high-value crops, cotton, sorghum and other field crops. Cooperating farmers also test innovative marketing techniques.

Our Mid-Atlantic Network, started in 1989, investigates on-farm composting methods and economics, rural-urban marketing opportunities and development pressures, farmland preservation and other issues. We began a Metro-Farmer Networks Program in '92, to encourage the adoption of innovative marketing methods, reduce production constraints and promote sustainable practices on the "urban fringe."

We're encouraged by what we see as a ripple effect that these and other sustainable agriculture networks are having on American farming. Many forward-thinking individuals and private foundations share our interest in promoting regenerative farming methods. With their help and yours, farming in the 21st century will enhance our resource base, rather than merely maintain it or drain it.

Consider this book a high-powered yet very affordable tool for improving *your* on-farm resources, starting right now. "Regenerative methods change natural environments for the better," Bob Rodale said. "The practitioners of regenerative technology aim high. They look well beyond the realities of today to a world that can no longer afford the free use of non-renewable resources."

Welcome to farming's future.

A Model Of Diversification

Rodale Midwest Network cooperator Terry Holsapple raises pumpkins, grain and other crops on his farm in Greenup, Ill. He plans to set up paddocks for rotational grazing of cattle soon.

Vetch And Vegetables Add Value

Cover crops and diversity reduce risk and expense.

GREENUP, Ill. — Unlike many farmers who worried about maintaining cash flow in the turbulent '80s, Terry Holsapple and his wife, Sheila, *reduced* their acreage to increase income. They went from farming 900 acres in the mid-'70s to a more manageable 600 in '84.

Cover crops, especially hairy vetch, allowed them to phase out synthetic fertilizer and herbicide on their corn, soybeans and wheat. And high-value crops such as pumpkins and melons have boosted gross income to as much as $2,000 per acre on some ground.

The Holsapples have 120 certified-organic acres on their south-central Illinois farm. Terry, 38, also helps his father farm another 480 acres with reduced inputs.

Holsapple started shaving purchased inputs in spring '80, when he cut his anhydrous rate by one-third (to 100 pounds of N) on a 13-acre field where corn followed a 3-year-old alfalfa stand. Cutting the N rate for corn after a legume plowdown seemed logical at the time. "We'd been seeing tremendous yields after alfalfa. And before people got away from rotations, my father had been lowering his N rates after alfalfa and clovers," says Holsapple.

Terry and Sheila Holsapple
Greenup, Ill.

Started cutting inputs: 1980
200 acres corn
200 acres soybeans
40 acres wheat
50 acres hairy vetch seed crop
40 acres alfalfa and other hay crops
40 acres pasture
30 acres vegetables
600 acres (120 certified organic)

Growing Season: 178 frost-free days
40 inches of precipitation

"We grew 160-bushel corn on that ground, when across the road, where corn followed soybeans, we could only get 100, maybe 110 bushels."

At the same reduced rate of nitrogen on a rented tract, corn after red clover yielded 150 bushels, despite some lodging.

Reluctant to cut inputs drastically, Holsapple used normal fertilizer and herbicide rates on the soybeans he planted in '81 on the former alfalfa field. Despite the use of a variety prone to lodging in wet weather, the field yielded 40-bushel beans, 5 bushels more than his usual harvest. It made him wonder what his yields could have been had he grown a better variety of soybeans.

"In the early '80s, times were tough for everybody," says Holsapple. "Costs were going up, prices going down, and interest rates were approaching 20 percent. Cost control was a necessity. It didn't matter what you thought about inputs. You had to do something."

In corn in '82, again with just 100 pounds of N per acre, the field really turned around, he says. "It made 180-bushel corn, which you just don't get very often in this county."

Cashing In On Vetch

Holsapple cites two other reasons why '82 proved a pivotal year: He started reading *The New Farm* magazine, and he got hooked on hairy vetch. A story about Glenn and Rex Spray of Mount Vernon, Ohio, who were raising 450 acres of organic corn and soybeans ("These Farmers Pocket Profits And Save Soil," *The New Farm*, March/April '82), convinced Holsapple that lower production costs were possible without losing yields.

An article about Delaware farmer Bill Grantham's overseeding hairy vetch into corn to provide nitrogen "boggled my mind," he says. "I'd been educated in a chemical-intensive, high-fertility program. So I was skeptical, especially about the vetch. I had to try it."

In fall '82, he planted 3 acres of hairy vetch on set-aside, "mostly on end rows and on river-bottom ground that was foul with weeds and wouldn't grow anything else. Well, the vetch came on like gangbusters. Putting on luxurious growth in spring '83, it reached a tremendous height by the middle of May."

He decided to try using hairy vetch for nitrogen before corn, and to combine some for seed. He broadcast vetch at 10 pounds per acre (less than half the recommended rate) into 30 acres of soybean stubble in fall '83. A cold, wet spring stunted the vetch's early growth, however, and gave a poor stand on two-thirds of the acres. So after plow-down, Holsapple applied 45 pounds of N per acre at corn planting, using urea-ammonium nitrate. At just a third of his normal N rate, "the corn did really well."

Besides cutting his N rates, Holsapple started trimming herbicide rates and scaling back on P and K as well. He soon had his cash expenses for corn down to less than $30 per acre in some fields, compared with $75 per acre in the '70s.

In '86, his corn yielded 156 bushels on one 12-acre field. That was with just 18 pounds of added N (100 pounds of 18-46-0) per acre after plowing down a waist-high stand of vetch.

"I was still using some herbicide then, but I was having a custom applicator put it on at one-quarter the rate, inducting it with the fertilizer. From that point on, though, my home acres never received *any* synthetic inputs.

"Whenever we've raised corn after vetch, yields have been as good as or better than chemical corn. I don't know why anyone uses chemicals on corn now. Corn is extremely easy to keep weed-free without herbicides.

"I haven't seen any decrease at all in soybean yields, either, since cutting inputs. Not even the first couple years.

"I didn't like growing beans back in the '70s," admits Holsapple, who grows about 200 acres of soybeans each year. "You fought the grass, you fought the weeds. Soybeans seemed so risky. But I learned how to control weeds mechanically in beans during the '80s. They're more sensitive to the timing of cultivation practices than corn is."

Good Stands Essential

Use of an inoculant each year is important for a good vetch crop, Holsapple found. "Without it, we'd see just a fourth or even a sixth of the stand we expected. The inoculant is cheap insurance at about $1.25 per bushel of seed. That's about 60 cents per acre."

He hasn't seen any evidence of soybean cyst nematodes in beans following vetch, which is an alternate host of the pest. But he's not growing vetch in the summer, when any nematodes might be active, he notes.

A 13-acre field of hairy vetch astounded Illinois farmer Terry Holsapple by producing more than 600 pounds of seed per acre in spring '92.

"The vetch planting date is critical. In southern Illinois, the standard way is to drill 25 pounds per acre after August 15, with September 1 being prime. We plant a bushel of oats with it as a nurse crop to allow an additional month of growth in fall. It comes on good, the oats die and in spring the vetch takes off.

"Another way is to cut wheat the first week in July, then go in immediately and put down vetch. Whether you drill or broadcast and disk it in, it'll get rank. The vetch will be knee-high by the end of October. But that's too tall and it will winterkill. You can disk it really light and leave the crowns. Then it survives and comes on in spring. But you've already got 150 to 200 pounds of N in the fall topgrowth, so it doesn't matter much what amount of spring growth you see.

"But I generally don't seed vetch right after wheat harvest. I've mowed some back before the winter instead of disking, though, and *that* works too.

"It's easy to get 200 to 250 pounds of N per acre from a vetch crop around here. Our best vetch stand provided 376 pounds of N per acre from the tops alone in spring '89. I don't have to worry much about nitrate leaching, either, as vetch provides a slow release over four years with about a 50-percent flush the first year, depending on weather."

By keeping his cover-crop establishment costs to less than $23 an acre — $10 for 40 pounds of home-grown vetch seed, $3 for 1 bushel of oats and $10 for disking and planting — Holsapple estimates he gets available N for about 18 cents a pound if he assumes 50-percent availability of N from vetch the first year. Other vetch benefits include weed control, erosion prevention, reduced compaction and better soil structure — and they're practically free, he stresses.

Flexible Rotation

Holsapple has never had a regular rotation. "I want to be as flexible as possible," he says. "I usually plant soybeans after a cover crop such as vetch or rye, then follow the beans with either fall-planted wheat or vetch. Whether it's a wet or dry spring, I have a lot of options. If I have vetch still in by June, for example, I can harvest the seed or cut it for hay. With wheat, I have the option of disking and planting beans, cutting for hay or harvesting cash grain or seed.

"I don't want to feel compelled to plant corn or soybeans in a given field. And I never plant corn after corn, or soybeans after soybeans. On my organic acres now, I tend to grow as many winter-annual crops as possible, such as wheat, rye or vetch, rather than corn."

When he harvests hairy vetch seed, Holsapple finds it easy to sell the 350-plus pounds per acre to neighboring farmers at 60 cents per pound. That's at least $210 gross income per acre, not to mention the biomass left in the field.

During the '88 drought, 60 acres of a rye/vetch mix seeded the previous fall yielded 400 pounds of vetch seed per acre, plus 20 bushels of rye per acre. "The vetch seed alone made more money than most of the drought-stressed corn around here. The rye was gravy," he recalls.

"I sell vetch seed for 15 to 20 cents cheaper per pound than other suppliers. But I'm not out to be a big seed seller. I want to promote cover crops. I tell folks to grow their own seed if possible."

Holsapple has been saving his hairy vetch seed each season since '87, so it's well-adapted to his growing conditions. His vetch produces mammoth seed pods with more seeds per pod than many varieties. "Dr. Marianne Sarrantonio [legume project coordinator at the Rodale Institute Research Center] says the hairier the vetch leaf, the more winter-hardy it is, and she says my vetch is the hairiest she's ever seen."

He didn't need any new equipment for cover cropping. "I'm still using an old John Deere grain drill, and our John Deere 7700 combine for our cover-crop seed harvesting. It works well for our vetch and should work for just about anything," Holsapple says.

He also likes the way cover crops reduce weeds in next year's row crops. He started phasing out herbicides for his corn and soybeans in the early '80s, and relied on two rotary hoeings and two cultivations to take out any weed survivors.

Holsapple already owned a six-row Noble cultivator with Danish tines plus an old John Deere row-crop cultivator with half sweeps. He borrowed a rotary hoe from an uncle, until finally buying his own 30-foot John Deere hoe in '88.

"I like to hoe all my row crops the fourth and eighth days after planting," says Holsapple. "If you want to get rid of herbicides completely before you hone your rotation, you need a rotary hoe. But it's not a cure-all. It's understanding how to put together your whole low-input system that matters: rotation, cover cropping, rotary hoeing, cultivation and timing.

"My soil-test values have not dropped one iota since '79. If anything, potassium seems to be increasing. Last year, though, I saw a 100-point K difference when comparing a June 1 sample with one taken on June 20. That field yielded 38.5 bushels of beans per acre, so soil testing still seems

Photo by Mitch Mandel, Rodale Stock Images

Watermelons grow really well following a vetch cover crop, says Holsapple, who usually retails most of his melons locally.

like an inaccurate science to me. I figure soil testing during mid-May to mid-June gives a pretty good picture of soil values, though, as that's when nutrient release is at a high point."

High-Value Crops Help

Although happy with his cost-cutting and cover-cropping progress in the mid-'80s, Holsapple figured his bottom line could be even better. He'd cut fertilizer bills in half, brought his herbicide costs below $5 an acre and eliminated all insecticides. "We were spending less, but still thought the cash-grain operation wasn't doing well enough. So we started scaling back and looking at what else we could grow and market," he recalls.

He cut his acreage by a third and diversified with high-value vegetables. The ease with which he sold 300 dozen ears of sweet corn from the back of his pickup one day in '83 convinced him there was a strong local demand for fresh-picked produce.

"I knew I could grow broccoli and cauliflower for local markets, as they weren't being grown in the area. And I started small — with just 500 plants of each in '84 — because you can lose your tail if you don't know what you're doing," says Holsapple.

That first crop was barely enough to satisfy his family and friends, and profits were skimpy. "You can't expect to make any money on high-value crops the first two or three years, because you'll be plowing everything back into the operation."

In '85, he expanded to several thousand broccoli and cauliflower plants, plus some sweet corn and 2 acres of U-Pick pumpkins. By '86, he was grossing about $3,000 total from the two cole crops, and four times that from 20 acres of sweet corn, 4 acres of pumpkins and 200 tomato plants.

Holsapple, who prefers to sell produce at the peak of freshness, found that harvesting the cole crops interfered with some of the field-crop work. "I also realized that people don't want to eat more than one serving of broccoli or cauliflower per week, so the market was limited. That's not the case with melons." He dropped broccoli and cauliflower in '87, and replaced them with muskmelons and watermelons.

He starts seven different melon varieties in a 13- by 60-foot greenhouse he built with windows salvaged from an old school. "We plant melons as soon as possible in the year. We had been using black plastic to warm the soil and prevent weeds, but the materials and labor were too costly." Now, he disks vetch two or three times before putting in transplants. "You can afford to make that many passes with small acreages of high-value crops," he notes.

After transplanting melons, he has the option of rotary hoeing once or cultivating. "If I set the hoe light and be careful, the plants take it well — I seldom rip one out," he says.

He cultivates melons with a one-row, 3-point-hitch tractor scratcher. He also uses a two-shanked "wiggle hoe" he made in '91. "It's just a garden cultivator I split in half and attached to the transplanter where the water wheel used to be. I welded brackets onto the transplanter, and a universal joint allows the hoe to move easily in any direction," he says.

The wiggle hoe and tractor scratcher, plus some handhoeing, take care of most weeds. "Even if we see a late flush of weeds, the melons are already made," he adds.

If it's a dry year, Holsapple puts down 3/4-inch, 8-mil disposable T-Tape drip irrigation hose to add water close to the plants. (Contact: T-Systems International Corp., 9650 Camino Breeze, San Diego CA 92126, (619) 578-1860.) In '91, he irrigated about 1.5 acres of melons. "I came within a day of putting it into the pumpkins too, but a 4-inch rain in July let us avoid that."

The low-pressure tape uses only 17 gallons of well water per acre per minute, so Holsapple's irrigation expenses are essentially the cost of the tape (about $175 per acre). "The tape's a pretty good insurance policy to have around."

"We've cut back on melon production in the past couple years, though. I can sell them, but I don't like selling into the wholesale market. That defeats the purpose. It's the same with labor. We hire some local high school kids as needed, but not many. I try to keep the high-value work in the family."

Insects haven't been a problem, says Holsapple, who rotates crops to minimize pest populations. And he hasn't used any botanical insecticides since '88.

For fertility and weed control in pumpkin fields, he relies on hairy vetch and mechanical cultivation. He disks vetch, levels it off and lets it lie for about three weeks. During that time, he field cultivates twice, with the Danish tines set an inch deep to keep out surface weeds. "Then I'll go in and plant with the corn planter, hoe twice, cultivate once, and that's it until picking," he says.

In a typical year, the Holsapples have 30 acres of organic high-value crops, mainly pumpkins, sweet corn and watermelons, with a few muskmelons, some gourds and Indian corn. The vegetables and fruit provide 50 percent of the farm's gross income.

Each fall, their on-farm harvest market attracts a big crowd. Haunted hay rides, homemade Halloween crafts, U-Pick pumpkins and a 3,000-bale hay maze make for a fun-filled family outing, Terry says.

A Learning Curve

Mastering low-input systems is an educational process that takes time, says Holsapple. "Information transfer has been my biggest obstacle, and that's probably still the case today. That's because most of it's farmer to farmer.

Terry's sister Kathy adds another plant to the quarter-acre black raspberry patch she and her mom manage organically. To control weeds, they use hay mulch and occasional mowing — that's Terry's father helping out.

"If I knew in the early '80s what I know now, I could've eliminated chemicals in about two years. But it involves a tremendous amount of reading and paperwork and researching articles. I'm still learning. That reading hasn't come down much."

Don't expect many shortcuts, says Holsapple. "As you turn to low-input farming, you improve your management and planning skills. It takes more long-range planning. You need to master your cover-crop timing, harvesting and other things. And you're monitoring the fields more. That's my favorite pastime — taking off and walking the farm."

If you want to reduce inputs, find a spot in your rotation that lets you begin with minimal risk, Holsapple says. "I suggest starting with wheat after soybeans. Wheat or oats give you a great place to start with fall covers. That lets you observe cover growth and ease into covers. It's almost too late in the season to start with vetch after soybeans.

Because her dad uses organic methods, 4-year-old Paige Holsapple says its OK to catch her in the rye.

"Then hit the books, talk to other farmers, attend meetings and understand how others are succeeding at this. Talk to others who have cut their chemical rates before you cut yours. You have to know it can work before you can make it work on your farm."

When you first try vetch before corn, you might want to use 40 to 60 pounds of sidedress in the corn, he says. "In a dry year, that's just gonna cost you, though, so don't worry about yield. After the corn, come in with rye, soybeans or wheat. As you add the organic matter, there'll be enough residual N that after a good vetch crop you shouldn't need any synthetic N on corn. You'll have the N cycle built in.

"Once I started, I soon eliminated half of my rates, worked it down to one-fourth, then knew I could go totally without synthetic inputs. That gradual transition also lets you see the soil working on its own.

"I don't have near the risk exposure with my field crops now, because I'm farming fewer acres and spending a lot less on each."

Still Changing

Always looking for ways to improve his field operations, Holsapple switched his primary tillage system to ridge till in '92. He bought a new Buffalo ridge-till planter and a used cultivator, plus an electronic cultivator-guidance system. He expects to recover his equipment costs within three years.

"I can use covers even more on ridges," he says. "It's a wonderful system. I've watched Iowa ridge-till farmer Dick Thompson's results for years. In addition to labor- and fuel savings plus better weed control, ridge till lets you do a better job of managing organic matter, he's shown.

"Working covers into your ridges gives you good biomass with less soil movement than in conventional tillage. You're not burning up as much organic matter *or* destroying earthworms. I figure I'll be able to maintain 120-bushel corn and 40-bushel beans at minimal cost."

Holsapple planted 160 apple trees in '92 and expects to pick his first crop of organic apples in '94. He put in eight different varieties that he purchased mail-order, based on maturity dates that will coincide with his pumpkin harvest — mid-September to late October.

He is venturing into rotational grazing too. "I started with 15 dairy bull calves to mow the lawn around the house last year, planning for a herd of 50 by '93. But prices looked good, so we sold them all. I'm improving the pastures near the house, though. I'm leaning toward stockers and maybe some brood cows once I've got the pastures up to snuff."

Whether it's high-value fruits and vegetables, seed production or livestock, diversification is the key, says Holsapple. "Having income every month means you have to borrow less. And as long as you don't put money out, you don't have to pay it back."

'Becoming Organic Wasn't The Goal'

Field by field, this dairy farmer eliminated chemicals with little yield loss.

LESLIE, Mich. — Instead of aiming for bin-busting yields, dairyman Bob Fogg, 40, has been profitably harvesting low-cost crops since '73 on 230 tillable acres in south-central Michigan. He and his wife, JoAnn, have also farmed 100 acres on his mother's farm since '84.

The Foggs are farming some of "the poorest soils in the county," or at least that's what their Extension agent said back in the mid-'80s. You wouldn't know it from their yields.

Thanks to manure from a 25-cow dairy herd plus the use of cover crops, the Foggs harvest corn, soybeans and small grains at or above county averages. And they're doing it *without* purchased inputs on most of the acres. Their home fields, many of which have been farmed organically since the mid-'80s, were all certified organic by '91.

Bob and JoAnn Fogg
Leslie, Mich.

Started cutting inputs: 1981
50-75 acres corn
50-75 acres soybeans
50 acres rye
50 acres oats
50-70 acres alfalfa
10 acres permanent pasture
2 acres vegetables
330 tillable acres

Livestock: 25-cow dairy herd with 25 heifers
Growing Season: 150 frost-free days
32 inches of precipitation

"We're not in hock over fertilizer," JoAnn says bluntly. "We don't ask for $5,000 to plant every year."

The Foggs started cutting inputs in '81, mostly for environmental reasons. "Becoming organic wasn't the goal. But it was an awful good feeling knowing we were moving away from chemical inputs, which aren't good in the long run for the land or for farmers," says Bob.

"We weren't really looking to save money. We just made one small change at a time, taking a field every year and seeing what fertilizers and herbicides we could get along without. Doing that helps your cash flow by reducing your need for input money in spring, of course. But you can't fully appreciate any cumulative cost benefits until after a few years of cutting chemicals."

To make the transition easier, Fogg reduced his corn acreage by more than 50 percent in '81, to about 40 acres. He started planting more alfalfa. He also plowed down an 18-acre field that had been in alfalfa for several years, and planted corn there without any synthetic fertilizer or manure.

"Since it was a small field with a good nitrogen credit from the alfalfa, I wasn't too worried about skipping fertilizer for the corn. I saved about $20 per acre by not sidedressing urea at 100 pounds of N per acre."

For weed control, he applied 1.5 pounds of atrazine per acre, and cultivated once. The field yielded the same as his conventional corn acres — more than 90 bushels per acre. "That's a lot better than I expected. It was beautiful — some of the best and cheapest corn I had ever raised," he says.

Manure's Cheap

In the same field the following year, Fogg again planted corn, but without *any* chemicals. He fertilized with 10 tons of dairy manure per acre and wasn't too surprised when he saw more weeds, especially quackgrass, velvetleaf and redroot pigweed. He cultivated twice, but yields declined to about 80 bushels per acre.

Was Fogg discouraged? Not at all. Without atrazine, "the corn was even cheaper to grow that year," he says. His dairy manure was free, he points out, and he didn't mind supplying the labor for hauling manure or for cultivating. And for corn after corn — not his usual practice — he wasn't expecting a better yield anyway.

He cut his '82 corn for silage, then broadcast rye. In spring '83, he frost seeded clover into the rye. As he expected, the chemical-free field was essentially weed-free. He harvested 25 bushels of rye per acre in summer '83 and plowed down the clover green manure in spring '84.

Fogg was already saving money by rarely using insecticides. He had sprayed Sevin for alfalfa weevil just once, in '75, but worried about the chemical's safety. A good rotation should eliminate the need for insecticides *and* purchased N, he figured. "Nitrogen is probably the most expensive input, yet it's one of the easiest items for a farmer to avoid buying. You can grow your own!"

By '84, Fogg had cut his cash outlay for a corn crop by 70 percent — from $100 per acre to just $30 per acre — without losing much yield. That year, he skipped his normal 60 pounds of nitrogen per acre for corn following clover. He spread 10 tons of dairy manure per acre on the 18-acre field, and also on an additional 24 acres of corn that followed alfalfa.

"Match the tractor power to the job," Bob Fogg tells field-day visitors.

On 7 of those 24 acres, he supplemented the manure with 100 pounds of 16-16-16 starter fertilizer per acre. The material was left over from one of his oat fields. The results surprised him. "You couldn't tell where I ran out of starter. All the corn looked the same. The grass and weeds were growing just as well on either side of the field, too," he jokes. Yields were identical, he recalls.

On the 18-acre field, however, he made the mistake of planting some 2-year-old seed from his corn crib rather than buying enough fresh hybrid seed. Germination was poor, and Fogg wound up spraying some 2,4-D just to keep weeds from overrunning the field.

"I hated to use 2,4-D, but in that case it made the difference between getting income from the field and losing the crop."

On that and other fields, the Foggs found that eliminating herbicides was the biggest obstacle between them and their goal of a reduced-input farm. "There were times when spending less than $5 per acre — including my own labor — for herbicide might have doubled my corn yield. But our field sizes aren't that large (most are 20 to 60 acres) and I was able to phase out chemicals on at least one field every year in the early '80s," says Bob.

By '86, all of their home fields had been "gobbled up" this way. That was also the last year Fogg used any herbicide.

"By then, I had already cut my atrazine rate in half, to just two-thirds of a pound per acre on the non-organic acres. Even when I used atrazine, which was my usual herbicide, I cultivated at least once. So I wasn't too concerned about the labor involved in mechanical weed control."

With Metal, Timeliness Counts

Fogg had been sharing a cultivator for many years with his father, who was raising poultry, corn and soybeans nearby. "It was getting to the point where I needed my own cultivator right now. Timeliness is really important when you aren't using herbicides."

So he bought a new four-row John Deere Danish-tine cultivator in '83. The cultivator's sweeps and rolling shields work in tandem to throw a small amount of soil beneath the crop leaves for some in-row weed control.

"I knew I could eliminate early weeds in the row better with a rotary hoe, so I bought a new John Deere rotary hoe in '86. Since then, weed control has been easy.

"It seems to me, though, that cover crops and proper rotations are even more important than cultivating or rotary hoeing when it comes to controlling weeds without herbicides. We really haven't had any weed problems to speak of on our home farm since '89. My biggest problem has been broadleaves, especially some humongous burdock in small-grain fields. I let plowing control it.

"One side of one field has been particularly bad with burdock. Even rye hasn't controlled that area. If need be, I'll ax them out with a pointed shovel.

"I find volunteer crops more problematic than weeds. If I plant buckwheat, for example, I now know to wait at least two years before planting corn, because volunteer buckwheat can reduce corn yield by 50 percent. Buckwheat's a great green manure if you get it plowed before seed set.

"I see some volunteer corn in my soybeans. I could avoid that with small grains between corn and soybeans, but I prefer to grow corn where I think I have the largest volume of nitrogen accumulated. So I usually plant corn after a legume cover crop such as alfalfa, red clover, sweetclover or hairy vetch."

Fogg likes using cover crops and has participated in many Rodale network studies involving covers. In one study, he compared the benefits of sweetclover (frost seeded at 20 pounds per acre in spring '90) with a mix of hairy vetch and rye (seeded at 2 bushels per acre in fall '90). In the test strips the following year, Fogg found no significant differences in late-spring soil nitrate levels, percent of N in corn leaf tissue, or corn yields.

"I still don't feel that getting the highest yields is the most important thing. I think it's getting your money back — and then some."

"The sweetclover performed better economically, because seed and establishment costs were lower than for vetch with rye," he points out. "The vetch/rye could prove more flexible, though, since it can be harvested for seed if weather conditions or other factors prevent corn planting."

For seeding covers, Fogg uses a PTO-driven broadcast seeder that he bought new in '84. "It's actually a fertilizer spreader but with proper calibration you can seed 15 to 20 acres an hour. It gives a 16- to 20-foot swath for small seeds and a 25-foot swath for larger seeds. I could have drilled alfalfa anytime I wanted a forage crop, but I also wanted to seed a variety of cover crops quickly, with less expense."

Back in the mid-'80s when he had two manure spreaders, two balers and five tractors, Fogg figured he was a little overtooled. He had more horsepower than you'd find on a typical organic farm. Now, he's down to one spreader, one baler and three frequently used tractors.

"I match the tractor power to the job, and I usually buy used equipment that has to perform well for a needed task. I haven't bought a *new* tractor since '75. Machinery's critical during haying, though, so I've always bought new haying equipment. I believe in maintaining top-of-the-line haying implements."

Quit Milking?

The Foggs almost phased out their manure supply in '85. Despite a rolling herd average of close to 18,000 pounds of milk per year in the early '80s — 2,000 pounds more than county average — they decided to quit dairying in December '84.

"We'd thought about it for awhile, and the Milk Diversion Program gave us a chance to be paid to quit," says Bob. Begun in January '84, the 15-month program gave participating dairy farmers $10 per hundredweight for up to 30 percent of the milk they didn't produce, based on '81 and '82 production levels.

As cash-grain farmers instead of feed producers, the Foggs knew that minimizing their crop-production costs was even more important than when they were milking. "I still don't feel that getting the highest yields is the most important thing. I think it's getting your money back — and then some," says Bob.

He knew he'd miss the free dairy manure, but figured he would work in more cover crops and green manures to replace it. He was happy with the progress he was making at building natural nitrogen reserves in his fields. But he was still giving most of his '85 corn crop 60 pounds of purchased N per acre and three-fourths pound of atrazine per acre.

Three cornfields (100 acres in all) that year yielded an average of 133 bushels per acre. That was 30 bushels more than normal for his farm and 15 bushels higher than county average.

One of the cornfields, a 19-acre tract following a 5-year alfalfa stand, received no purchased N at all. With just 6 tons of dairy manure per acre (applied the previous fall and winter), three-fourths pound of atrazine per acre and a single cultivation on most of

With milk prices low and plenty to do, JoAnn and Bob might sell their cows — again.

the acres, the field yielded 139-bushel corn at a cash cost of just 34 cents a bushel.

To feed the heifers he hadn't sold, Fogg also planted 14 acres of corn for silage. That field, which followed a 3-year alfalfa stand, received no purchased inputs and yielded more than 15 tons of silage per acre. And 42 acres of chemical-free wheat yielded an impressive 55 bushels per acre.

Dairying Again

"Crop prices were good in '84, but in '85 corn prices were dropping," Fogg recalls. "We'd kept all the heifers and enough winter feed for them, so we soon had to decide if we'd sell them or keep them for dairying. We started milking again in October '86. Since then, we've maintained about half the herd size as before: 25 cows and about that many heifers.

"I remodeled the barn before milking again, to give the animals additional space and more comfortable conditions. I put in tie stalls in place of the stanchions I had.

"I still lack outside storage for manure. I can pile for two or three weeks in the barnyard, but I prefer to spread manure almost every day. I usually have at least three different fields I can spread on, depending on the field conditions — even in winter. I don't want the nutrients running away, though, plus there's a creek nearby so I have to keep that in mind. So I usually have a rye/vetch field or an alfalfa field that I can spread the manure on."

Fogg needs at least 40 acres of alfalfa for hay. He also feeds dry shell corn and a mix of oats, barley and roasted soybeans.

At least half of the Foggs' 50 to 75 acres of corn each year is intended for livestock feed, with the balance available for the organic grain market. With up to 100 acres of small grains each year, Bob invariably grows some rye and oats. Depending on his feed inventory and field availability, he'll occasionally plant wheat and barley.

"I usually have soybean screenings left over from previous crops, but I ran out in early '92. I purchased some protein supplement for the rest of the winter, for the first time in 10 years.

"I watch protein levels closely in winter. They have been a little lower in the last two years because I've been 10 to 14 days late with my haying both years. That has probably cost me some milk production. If I get my hay schedule back on track, I really wouldn't need any protein buys."

Labor has been a limiting factor for Fogg, even before he cut chemicals. He has more acres than he can plant to row crops each year, but hopes his oldest son, Nathan, 16, will soon be able to help with rotary hoeing and cultivating.

Even so, he might phase out the cows again in the coming years. "They're a lot of work, and with low dairy prices, they don't bring in as much money as they used to. If I sell the cows, I'll have more time to devote to the crops and marketing, and I might acquire some other livestock. I could also expand the vegetable operation (currently less than 2 acres), especially if the kids take an interest." Fogg has been test-marketing squash and other vegetables at organic cooperatives in and near Lansing, Mich.

Marketing Organic Grains

Completely organic on the home farm since '87, the Foggs sold their first certified-organic crops in '89. "Because we've always kept detailed, field-by-field records, it wasn't difficult to get certified by the Organic Crop Improvement Association (OCIA). Selling organic takes special effort, though, and I sometimes wonder if it's worth it," says Bob, who sells some organic soybeans to Eden Foods Inc., Clinton, Mich.

"We often sell our oats, soft winter wheat and barley on the conventional market, or feed them to our cows. But we'll keep *growing* our crops organically. We take the marketing a day at a time. When you're on the phone selling crops, you can't be out in the field.

"Growing organically gets easier every year you're away from chemicals. Early on, we lost some yield in a few fields. But we learned from our mistakes. Now I have the confidence that I can control weeds without herbicides. With cover crops and hay to plow down, nitrogen is no problem, and phosphorus levels are holding steady. Some fields have lower potassium soil-test levels than I'm accustomed to, but so far it hasn't hurt yields."

Fogg has every field tested about every three years. "I see values all over the board at times, usually with no correlation with yields. We can bring low values up gradually — with manure and cover crops — and the high ones don't seem to be dropping.

"Our organic-matter levels are creeping up. A lot of them test at 2.5 percent now."

Fogg says the labor and management has increased a bit. "But you learn how to work around it. You make time for whatever's important. Your outlook influences your planning. It's not as though you have to schedule your planning, though. I do it when I'm milking or doing the chores."

Like most Rodale network cooperators, Fogg has found that when you cut chemicals *successfully*, your public image changes. Local acceptance of his farming practices has increased. "None of my neighbors think I'm a whack anymore. And I get asked to do more speaking engagements now," he says.

The Foggs' annual field day consistently attracts more than 80 people. The farm has also become a popular stop on SCS and other farm tours.

Go Slowly

Give yourself a reasonable time period for a transition from purchased inputs, Fogg suggests. "Don't get bent out of shape if something doesn't work out the first time. Don't expect dramatic results the first three to five years. Do it on a 10-year plan and go slowly. Take a few more risks and experiment. Set one field aside at a time and maybe eliminate purchased fertilizers.

"Covers allow some nitrogen cutback plus soil benefits that let you get away from chemical fertilizers. You still need the nutrients from somewhere, though. I've been getting them from cover crops, manure, the soil and the atmosphere. You need healthy soil, and you need efficiency."

Will it pay off? "I wonder about the financial difference between conventional and organic farming sometimes. With my small field sizes, I could take chances in the transition process without risking too much income. I know we're better off environmentally and healthwise, and we're more self-sufficient than before. We think we're better off financially too. But in terms of net farm income, I can't really gauge what our income difference has been since we've converted to organic practices.

Handling and marketing organic grains and soybeans takes time, but Fogg finds the extra effort rewarding.

"The biggest difference to me is the satisfaction of knowing you're doing what you like to do without destroying the activity you like. And you're helping the planet. I'm happy to have been able to do what I've accomplished with fewer chemicals the last 10 years and sure want to encourage others to do the same."

'I Learned The Hard Way'

You still need nitrogen, this hog farmer found.

MADISON, Minn. — Carmen and Sally Fernholz became organic farmers almost overnight. That was in '73. Thanks to feeder hogs and smart financial management, they're still farming the original "home 80" organically. Plus, they're nearly chemical-free on another 220 acres, including 90 acres they bought in '91.

A 60-sow farrow-to-finish operation provides the manure that lets the Fernholzes grow corn, soybeans and small grains without synthetic nitrogen. Keeping their debt load to a minimum helped the couple weather three transition years of dramatically reduced crop yields, without risking the farm.

Land stewardship and an open mind about innovative practices come naturally to Carmen Fernholz, whose conservation-minded father was one of the first western Minnesota farmers to use a chisel plow back in the '40s. Although Carmen grew up on the family farm, it wasn't until '72 that the former English teacher and his wife decided to start farming on their own.

Carmen and Sally Fernholz
Madison, Minn.

Started cutting inputs: 1973
100 acres corn
60 acres soybeans
50 acres oats
50 acres wheat
15 acres alfalfa
2 acres carrots and potatoes
23 acres CRP waterways, terraces, grass strips and buildings
300 acres

Livestock: 60-sow farrow-to-finish operation
Growing Season: 120 frost-free days
25 inches of precipitation

They raised 80 acres of conventional corn and soybeans that year, using county-standard rates of fertilizer and herbicide. With some carryover nitrogen probably helping — the previous operator had used plenty of fertilizer — Carmen harvested a respectable 90 bushels of corn per acre and 35-bushel beans his first year.

Crop prices were strong in the early '70s, with corn bringing about $3 to $3.50 per bushel, soybeans $10 to $12 per bushel and wheat from $5 to $6 per bushel. Even so, Fernholz decided to reduce his use of synthetic inputs as quickly as possible.

"You have to sell for a profit. I figured if you rotated crops properly, you shouldn't need the synthetic inputs. And I wanted to get away from chemicals. It's too easy to substitute chemicals for management," he says.

"Cutting back on fertilizer or herbicide forces a farmer to depend on his own skills, to know what he can do for himself, and what he can and can't tolerate."

Aside from better economics and greater self-reliance, the Fernholzes also wanted to safeguard their health and that of their four children, who now range in age from 12 to 22. Concern for the environment also figured in.

"I eliminated just about every chemical on all acres in '73, especially nitrogen. In their place, I tried using an organic fertilizer that was high in trace minerals, for about $15 per acre. I was under the impression that the natural nitrogen cycle — from the breakdown of crop residue — would supply my N needs.

"I didn't have any livestock then and went cold turkey and I learned the hard way."

Corn yields plummeted, from 90 bushels per acre to 55. Although Carmen's soybean yields stayed at county average (about 30 to 35 bushels per acre), his 15 acres of wheat yielded a paltry 25 bushels per acre. All told, yield losses reduced potential income by about $5,000 that year.

Fortunately, Fernholz had plenty of patience and little overhead. The acreage was small, and a low mortgage made it possible for him to rent 120 acres from an uncle starting in 1974.

He shared equipment with his brother, Chuck, who farms 250 acres conventionally. "Chuck would put in two-thirds of the cost of needed equipment and I'd put in the rest. Plus, Sally was and still is

Fernholz designed a square-bale hay feeder to provide alfalfa for his hogs.

working full time off the farm.

"We were just excited to be out here farming. And prices were good. I sold my first wheat crop for $6 a bushel in '73."

What did the neighbors think of his organic methods? "Nobody criticized me, at least not directly," says Carmen. "I never thought to worry about it. I was farming in the area where I grew up. Once in a while, the instructor for the adult farm program I was in would say people were wondering what that crazy guy was doing east of town.

Hogs To The Rescue

"Of course, I was comparing my fields to my brother's. I saw more vigorous growth in his cornfields than in mine, almost from mid-season on. My small grains would start out quite well, too, but then stall a month or two later. That got my wheels turning. I figured it took time to get the N cycle going, but I wondered if I needed a nitrogen source."

In summer '73, Fernholz bought some pigs from his brother and started a hog-finishing operation in an empty building on his father's farm. The power of hog manure to bolster yields was an eye-opener.

"The manure gave me a tremendous yield effect right to the line where I spread it. I got 80- to 100-bushel corn and 40- to 45-bushel beans. That was a real awakening, and one of the biggest lessons I ever learned. That's when I realized the limiting effects of a shortage of N."

Determined to avoid synthetic fertilizer, Fernholz established a 55-sow farrow-to-finish operation in late '75 and early '76. He figured the hogs would boost farm income as well as provide a good deal of the nitrogen he needed for acceptable yields at low cost. He was right on both counts.

In '76, he constructed his own hog building. He also put in a 28,000-gallon storage pit that he pumps out twice a year, in spring and fall. Having talked his brother into buying a 2,000-gallon honeywagon, Fernholz could now spread 4,000 gallons of hog manure (equivalent to 160 pounds of N) per acre on a third of his home acres each year. He still does, usually applying the manure to cornfields.

"Chuck lets me lease the honeywagon for about $10 per load, as long as I maintain it properly. We keep it on my farm most of the time, since I use it more. Although it came with injectors, we found after a year's use that we preferred spreading manure over the top and working it in with tillage equipment. That gives a much more even spread, with a little less power and a little more convenience."

Thistles A Challenge

By 1975, all but two fields on the home 80 had been certified organic. Carmen was feeding many of the crops to the hogs, though, and he wasn't expecting premium prices for organic crops.

"I got $13 a bushel the first time I sold organic soybeans, in '78. I take advantage of the organic market when I can, but it's not a stable one, so I don't count on it."

A heavy thistle population kept two of his fields from being certified organic until '78. "I've found Canada thistles hard to control in small-grain fields," he says. "I thought I could harvest wheat, oats or barley before thistles matured, but found out otherwise. So I backtracked and resorted to spot spraying Banvel on some fields. That's still the only herbicide I use and the only chemical that keeps us from certifying all our acreage as organic."

Fernholz prefers to control weeds through cultural and mechanical means. To control grasses such as foxtail, for example, he'll plant spring grains as early as soil conditions allow. He avoids short-stemmed grain varieties that tiller less than others, and he relies on two years of small grains to minimize weeds before planting corn.

He plants corn and soybeans one to two weeks later than most farmers in his area. The delay allows the soil to warm and lets him take out weeds with one or two diskings or field cultivations just before planting. He's careful not to work the seedbed more than 3 to 4 inches deep, to avoid bringing up new weed seeds.

"I used to disk in early spring to flush weeds, then field cultivate right before planting. But my own experience and some '50s research I discovered from the University of Minnesota convinced me it's better to wait," says Fernholz.

His 20-foot John Deere rotary hoe effectively controls germinating broadleaf weeds. He likes how the hoe lets him take out weeds fairly quickly.

"The hoe was a big investment in '79. Back then, $2,600 seemed steep, as people figured you'd still need chemicals. In my case, I was putting out for metal rather than chemicals.

"But it took me eight years to master rotary hoeing," he admits. "That's because you only get a few days a year to learn how to use it, and in different weather each year. Tips from other farmers, especially [Boone, Iowa, farmer] Dick Thompson, helped me a lot. Now, I feel very confident about rotary hoeing."

Legumes such as yellow-blossom sweetclover (shown), hairy vetch or NITRO alfalfa could supply all the nitrogen needs of small grains in Fernholz's rotation.

When Fernholz started farming, he used a rear-mounted cultivator so he could band herbicides. "After I abandoned chemicals, though, I wanted a front-mounted cultivator so I could watch the cultivation more closely, make finer settings and handle contours better. In '82, my brother and I finally bought a front-mounted six-row John Deere row-crop cultivator equipped with half-sweeps next to the row and full sweeps in the middle.

"I wanted to minimize weeds going to seed in the many point rows on my contours, so we got dual hydraulics. It lets me raise one side of the cultivator and still keep the other side in the ground.

"I was getting some weed misses, though, because I couldn't get the rear gangs mounted quite right. So I bought a neighbor's rear-mount four-row cultivator. I hitch it on the same tractor as the front-mount and use them both. I love it. For just $800, combined, they've been a tremendous investment. They eliminate weeds that survive hoeing and give me a fluffy top inch or two of soil that conserves moisture and keeps new weeds from sprouting."

Fernholz estimates he was saving at least $7.50 per acre during the '80s, thanks to rotary hoeing and cultivation in place of herbicides, and only occasional use of Banvel for Canada thistles. He also was saving about $30 per acre by not using synthetic fertilizer, and up to $20 per acre by avoiding insecticides.

"Any pest problems I see are ones typical of the whole region. Half of the state had a grasshopper problem in '89, for example. I came within days of spraying the cornfields, which had 20 to 25 hoppers per stalk of corn in July. But I received 2.5 inches of rain, which saved the crop. The corn hadn't silked yet. Cool, wet weather slowed the hoppers, then the corn pollinated and outpaced them.

Photo Courtesy of The Furrow

A Nimbus digital thermometer helps Fernholz monitor soil temperature and verify best planting dates for corn.

"I didn't spray an armyworm problem, either, in '90. There's only a narrow window of opportunity to spray for armyworm, and not all fields were infested. Everyone had 5- to 10-percent yield losses that year."

Flexible Rotations

Fernholz's soil-test values have been relatively stable, he reports. "That surprised me, though maybe it shouldn't. Manuring seems to fill the bill for P and K. And organic-matter levels have slowly risen, to about 3.2 percent."

On the organic acres, he would like to cut his manuring rate in half, to just 2,000 gallons per acre. That would let him spread more manure on the other acres. Before doing that, he thinks he should fine-tune his use of winter-hardy legume cover crops.

Most farmers in Fernholz's area rotate corn, soybeans and wheat. For more weed control and livestock feed, Fernholz normally adds an *extra* year of small grains, usually oats. He also likes to include a legume green manure, such as hairy vetch, in the wheat year.

"For $300 in '85, I bought a Herd Cyclone grass seeder that proved itself over the years with vetches and other legumes. But lately I've seen even better results planting winter-hardy legumes with an old endwheel drill behind my chisel plow when tilling grain stubble in fall. Compared with interseeding cover crops with small grains in spring and then leaving the covers until the following spring, fall tillage seems better at keeping weeds and grasses from setting seed."

Fernholz was surprised that hairy vetch survives Minnesota winters. "After fall seeding 20 acres with vetch in '91, I saw well over 90 percent survival in spring '92," he says.

He has found that small field sizes plus his four-year rotation give him the flexibility to put together a balanced hog ration, primarily from homegrown grain. It also helps him cope with ever-changing provisions of federal commodity programs.

"Because it's a confinement hog operation, I have no need for pasture or hay. To realize some of the benefits forages contribute to cropping systems, though, I've been experimenting with seeding different rates of nondormant alfalfas — especially NITRO — with small grains such as oats at different planting times. I'm also experimenting with feeding alfalfa to the hogs.

"I've been trying the NITRO on research plots and about 10 acres of set-aside, interseeded with oats, wheat, rye or barley for a double crop. I've found that NITRO doesn't tolerate nurse-crop pressure that well, though. You have to plant oats or wheat thinly when they're grown with NITRO. It looks like 15 pounds of NITRO seed per acre is best if you're planning to harvest the alfalfa and want good coverage and weed control."

Even in fall, NITRO provides more nitrogen from its roots than it does from its topgrowth. "We're looking at 80 to 100 pounds of nitrogen being fixed by NITRO, with about 60 percent of that from the roots. With a decent stand, I can bale it and feed it free-choice to my hogs and sell any extra. Not only does it supply ground cover and help control weeds, but it could also provide extra N for my grain crops.

"A major thrust of mine since the '70s has been to enhance the land's natural N, using manure, crop rotations, legumes and cover crops, municipal leaves — anything that improves the natural N cycle."

Fernholz started using leaves after town officials banned leaf burning in Madison in 1975. Fernholz convinced them to stockpile and haul some municipal leaves to a different field of his each year. "It's important not to incorporate the leaves immediately, as buried leaves tie up large amounts of soil nitrogen while decomposing quickly. The leaves rot slowly if you spread them out adequately. Aside from cooling the soil like a mulch, the leaves won't benefit the following summer's crops that much. But once fully decomposed they provide some good long-term organic matter." He receives about 50 tons of leaves annually.

Starting in 1990 as part of a three-year grant from the Minnesota Department of Agriculture, Fernholz began a comparison of NITRO alfalfa, hog manure and urea as nitrogen sources in a small grain-corn-soybean rotation. Hot, dry weather in '90 plus a continuing grasshopper problem stressed many of the NITRO treatment strips, which received an average of just 46 pounds of N per acre from the NITRO.

In '91, a wet year, NITRO provided an average of just 79 pounds of N per acre (roots and topgrowth). Corn yields averaged 72 bushels per acre in the manure treatments that received an equivalent of just 46 pounds of N the *previous* year, 67 bushels per acre in the urea treatments receiving 60 pounds of purchased N in '91 and 45 pounds the previous year, and just 46 bushels per acre in the NITRO treatments. He's continuing to work with NITRO, plus a number of other cover crops.

"In the long run, I'd like to perfect a three-year rotation of corn, followed by small grain and a legume cover, then soybeans. The second year is where I can make strides, whether using NITRO alfalfa, hairy vetch, medics or clover. I also could interseed vetches into ripening beans in the third year."

With hog prices — and interest rates — declining in '91, Fernholz was ready to expand the cash-grain side of the operation. He also wanted to put in more border plantings and grass waterways, and enroll some erodible land in the Conservation Reserve Program. He purchased another 90 acres, bringing the farm's total to 300 acres.

"With the new land, I hope to be able to stick with a three-year rotation. Initially, I'll be applying some synthetic nitrogen on new corn ground, probably about 40 to 50 pounds of N per acre from urea.

"I can operate with minimal purchased inputs and get economically satisfactory yields in corn. Most conventional farmers in the area run $15 to $20 of herbicides per acre of corn and at least that much per acre for beans. I'll spend just $2 to $3 per acre for herbicides for corn and none in soybeans.

"Farming more acres will take a little more labor than before. But assuming average weather, I don't anticipate falling behind in fertility and weed management." He expects to continue to share equipment and some labor with his brother, he notes. "Chuck is still farming conventionally, but my crops are comparable to his in all respects.

"There were a lot of days in the early '70s when I really doubted what was going on, questioned my goals and what I wanted to accomplish. Weeds were a challenge at times. They still are. But then things started going right, and that has given me a real sense of accomplishment.

"Maybe I'm a slow learner, but it's taken me quite a few years to accept what nature has out there and to really understand how nature works. Some information comes with time and experience. For example, knowing when weeds will grow, when soil temperature is optimum for tillage and at what moisture level you can harvest crops without losing crop quality.

"I had been turning the fan off at night in my solar grain drying bin. Then in '89, I found out I could dry corn in 12 to 15 days by not turning the fan off. I can harvest corn at a 30-percent moisture level and know that I'm not going to have a problem."

Fernholz is well-known for his work promoting sustainable farming. He serves in leadership positions in several organizations, including the Organic Growers and Buyers Association and the Minnesota Institute for Sustainable Agriculture. He's a member of the Midwest Sustainable Agriculture Working Group and speaks frequently at regional and national meetings. He's been featured in Land Stewardship Project videos as well as in many newspaper and magazine stories, including the January '92 issue of *The Furrow*.

Fernholz offers four tips for others who might want to reduce their use of off-farm inputs. "First, take no more than 10 to 20 percent of your acreage and commit to at least five years of reduced inputs on those acres. Start with something that isn't going to break you, so you can experiment and expand on it.

"Second, keep good notes. Third, read a lot, go to workshops and pick the brains of those who've tried things.

"Finally, be aware that your management styles are unique to you. As Dick Thompson says, 'Adapt, don't adopt.' So fit others' techniques into your style. When you make it your system, you'll see it work."

Who Says Cotton And Chemicals Are King?

Diverse organic crops and multiple markets rule this Southern farm.

TIPTONVILLE, Tenn. — When Lorna McMahon became fifth-generation owner of 2,300 acres of prime Mississippi-Delta farmland in '87, her northwest Tennessee neighbors soon took note. McMahon, who had run a scuba business in South Florida, started managing 230 of the acres herself — and organically at that.

The uproar among the neighbors was incredible, she says. "It was bad enough my being female, a major landowner and a farmer without an agronomy background. But to be organic was just more than the guys around here could handle. We had lots of 'doubting Thomases' who were sure we would fail. But I thrive on people telling me something can't be done."

Rising to the challenge of what she calls an "intense learning experience," McMahon sought and received advice from organic farmers across the country. She hired custom operators for planting, cultivating and harvesting her field crops, all grown without chemicals from the day she took over. As another of her priorities, she sought and secured markets for almost all of her organic crops, including cotton, soybeans, corn, pecans and vegetables.

Lorna McMahon
Tiptonville, Tenn.

Started cutting inputs: 1988
30 acres corn
50 acres soybeans
30 acres wheat
66 acres cotton
14 acres high-value crops
40 acres cover crops
230 acres; leases out 2,070 acres to other farmers

Growing Season: 215 frost-free days
48 inches of precipitation

"The neighbors went nuts when they found out we had a market for our organic cotton. They had the Extension agents in to see if we were going to be sending boll weevils their way. But insect counts showed that our predators outnumbered pests about 22 to 1," says McMahon.

"I have a contract for a million pounds of organic cotton in '92, and all those doubting Thomases are showing up on our doorstep. We're getting twice the money they are."

Working With Nature

"We just haven't had problems with alternative farming practices," McMahon says. "Nature and the seasons just work for us. It's not luck. It's proving what we've known all along: Sustainable agriculture works! You don't have to use all those chemicals."

Until he retired in the late '70s and rented out most of the farm, McMahon's father had raised cotton, corn, soybeans, wheat and milo, along with a few hogs and cattle. His *farming* methods were conventional, but he gardened organically for the family table. Large-scale organic practices, he told Lorna, would have to wait a generation. He didn't think there was a market for organic food, although he felt pesticide-free practices made sense.

"My dad died from cancer, firmly believing it was caused by farm chemicals. For health reasons, my husband John and I decided we wanted to reduce the use of chemicals in our farming," says McMahon.

Their first step in fall '87 was to start cleaning up the perimeters of the fields they intended to farm the following spring. With Extension and SCS advice, they began improving tree rows and borders to attract more wildlife. They started planting 60- to 70-foot-wide grass strips around field perimeters to filter any water draining into ditches. They also removed piles of trash; planted 40 of the acres to buckwheat each year starting in spring '88; and began composting cotton scraps, yard wastes and some greenchop buckwheat.

"We let buckwheat grow until flowering, but mow it before it seeds. We compost the first cutting, let the buckwheat flower again, then till it under," says McMahon.

"We decided to stop pesticide use entirely in our first year of farming," she adds. "We wanted a complete ecosystem on the farm. So I started putting together a program that I thought would work for us. It was pretty much trial and error, with a lot of farmer-to-farmer contact.

"The chemical farming had just been scratching the soil, I figured. We wanted to farm the soil deeper and thought crop rotation, green manures and some deeper tillage would help."

She started rotating soybeans, corn and wheat in fields where tenants had previously monocropped, especially for cotton. On the land she and her husband farmed, she avoided growing cotton entirely until '91.

Clean Fields, Respectable Yields

Even without synthetic fertilizer, yields remained stable the first few years. Corn still topped 90 bushels per acre, soybeans yielded more than 35 bushels per acre and wheat averaged 55 bushels.

"Our only real yield losses occurred in '91, but that was countywide. After a wet spring, a very dry summer had many people plowing under their wheat fields. The area's wheat suffered from a scab disease, and I wound up with one of the highest wheat yields in the county at just 27 bushels per acre," says McMahon.

That year, the McMahons grew 36 acres of cotton, which yielded a respectable 1.25 bales per acre. (Each cotton bale weighs about 500 pounds.) Although less than the county average of 2.0 bales per acre, it was impressive because it was achieved without any chemicals whatsoever. By comparison, the mean cotton yield in the southeastern United States is 1.3 bales per acre, according to USDA data. And 97 percent of cotton acres in the Southeast receive synthetic fertilizer, plus an average of 11 insecticide sprays and three herbicide treatments.

Weeds weren't much of a problem during the McMahons' transition from chemicals. In '88, they saw more Johnsongrass, but only in two or three fields. Moldboard plowing and two diskings followed by two crops of buckwheat helped in those fields. For all their field crops, the McMahons relied on custom cultivation rather than herbicides to take care of weeds. Most fields received two cultivations, and occasionally three.

Tennessee farmer Lorna McMahon and her 3-year-old, Hazen, enjoy the beauty of growing cotton organically: Besides selling the lint for the "O Wear" clothing line, the McMahons can market almost every part of the plant except the leaves.

Custom-operator expertise usually worked to everyone's advantage, they found. "In spring, it only takes about four or five days to get our fieldwork done. By custom hiring their equipment and labor to us, some of our neighbors get some cash flow at a time of year when they're cash-poor. We pay $25 per hour for a tractor and driver (for planting, disking and cultivating). They don't charge for travel time to the field," says McMahon.

"At harvest, we usually hire the same operators. We pay 50 cents a bushel in labor for wheat, soybeans and corn, and $4 per hundred pounds of gross lint for cotton ($20 per bale). You can't beat that, and we benefit too. It has let us diversify without spending too much money in expanding."

In spring '88, however, the farmer that McMahon hired for some custom tractor work applied Scepter for a flush of broadleaf weeds without checking with her. That delayed the organic certification of many acres for an additional year.

In '91, the cotton harvesters required 36-inch rows, which made it harder to cultivate the cotton. The McMahons spent about $42 per acre for cotton cultivation that year, which was $12 more per acre than they wanted to. In '92, they went with 30-inch rows in all crops and used a new custom operator for the cotton acres. "We lucked out. Only two people in the county have 30-inch-row cotton pickers, and one of them wants to grow some organic cotton," says McMahon.

> *"When we started farming, my husband and I thought people really needed organic products, for health and environmental reasons. With the organic movement building, the time was right."*

A cold, wet summer in '92 made weeds more of a problem, though. "Even the chemical growers had a lot of hand-weeding expense," says McMahon. In '91, she only spent $45 per acre for hand-weeding labor in her organic cotton. That expense soared to $100 per acre the next year. "By comparison, most conventional growers spend $50 to $75 per acre for herbicides, cultivation and hand-weeding," she says.

"We plan to buy our own four-row cultivator soon, so we can do some of the cultivation ourselves. It will help us be more timely with our weed control and it should pay for itself in one season with reductions in hand-weeding costs."

Diversity, Strip Cropping Ensure Fertility

The Tennessee Land Stewardship Association (TLSA) began certifying parts of the McMahon farm as organic in 1990. First certified was a 7-acre market garden that includes their 5,000-square-foot greenhouse. They use the parcel to grow everything from cut flowers and bulbs to exotic salad crops and open-pollinated heirloom vegetables. "We added two more greenhouses in '91 for an additional 5,000 square feet of growing space," says McMahon.

"I probably could have certified the market garden and a number of other fields earlier than '90, but we weren't really sure which previous tenants had used Scepter. I'm on the state certification board and I didn't think it right to certify any fields I was in doubt about. I thought it best to wait the full three years, plus the additional year where Scepter was custom sprayed by mistake in '88.

In '90, McMahon grew 50 acres of corn and about 170 acres of soybeans. She sold 1,000 bushels of the beans at a premium price to be processed into tofu. The McMahons also harvested good yields from 10 acres of pecan trees that year, which was the first year they marketed high-value products. The orchard's only fertilizer had come from some calves and horses that browsed the fescue understory.

The following year, McMahon and her husband certified part of the pecan orchard, along with 27.5 acres on which they double cropped wheat and soybeans. In her other fields, she grew 30 acres of corn, 36 acres of cotton and about 120 acres of soybeans. Yields were lower than expected countywide, because of adverse weather.

The McMahons enrolled 19.5 acres of grass field perimeters in the Conservation Reserve Program in '92. They receive a $65-per-acre annual payment to maintain the cover.

Also in '92, TLSA certified half of a 55-acre parcel on which the McMahons began strip cropping. They drilled 25-foot strips in a north-south pattern on most fields, alternating two cash-crop strips with one cover-crop strip. "We started with a buckwheat cover crop on the perimeter. We planted a strip of cotton or corn, then a strip of soybeans. Then we seeded the third strip with cover crops: a buckwheat/soybean mix in summer followed by a fall-seeded annual ryegrass/mammoth red clover mix. We repeated the sequence across the field, and will rotate the crops through each strip," says Lorna.

"We harvest both cover-crop mixes to add to the compost piles, then till in the regrowth. We'll probably add a winter cover crop between the cash-crop seasons too.

"That plus cover crops on some other fields amounts to almost 40 acres of non-cash covers on our 230 acres, mostly in strips. But that's the heart of our organic fertilization program for field crops. We wind up with a three- to six-year rotation of crops on the same parcel. That will reduce Johnsongrass weed pressure, raise organic matter, maintain a cover and continue to attract beneficial insects (when covers are in bloom) to our cotton crops."

Pests? What Pests?

The McMahons hire a scout for early detection of insects in their field crops. But their only pest problem so far has been some thrips in the cotton. Beneficial insects quickly reduced that threat.

"We haven't introduced *any* predator insects, though. They've all been volunteers," Lorna says proudly. "We were prepared to order about 600,000 green lacewings to battle thrips in '91. But lacewings moved in on their own, their levels increased and the cotton matured with minimal thrips damage.

"We also see some tarnished plant bugs in the cotton, but they haven't been a serious problem yet. We're going to have to find a beneficial insect or something organic we can use, though, in case we get an infestation."

In '91, the McMahon farm was one of only a handful in Tennessee to help in USDA trials of pheromone luresticks to control boll weevils in cotton fields. An insecticide coating on the stick kills the pests. "Although it's a poison, it's confined to the stick rather than the soil," says Lorna. "Our certifier decided it was a responsible way of handling boll weevils, so even certified-organic farmers in our area can use the luresticks."

Pheromone luresticks placed around field perimeters attract and kill boll weevils when they migrate between fields and overwintering sites. Pheromone-baited traps (inset) help McMahon monitor pest populations.

The luresticks are placed every 75 feet around the perimeter and are easy to identify. The sticks are effective for about 10 weeks, then McMahon returns them to the manufacturer for proper disposal.

Most conventional farmers spray insecticides to kill weevils when they defoliate cotton before harvest, she adds. McMahon has found that placing the pheromone luresticks around cotton fields in early spring helps reduce weevil levels considerably when the pests try to migrate into the fields. When any remaining weevils try to leave the field at harvest, most are killed by a second set of fresh sticks. McMahon also mows field borders to reduce the pest's overwintering habitat.

After inspecting McMahon's fields in August '91, a USDA entomologist told her that predator numbers were so high that she needn't be concerned about the few weevils present. "Neighbors don't have to worry about our boll weevils moving into their fields," says McMahon. "Instead, they should thank us for maintaining populations of beneficial insects."

Mild winters since '89 have led to a regional buildup in boll weevil populations. "In '94 or '95, we may be faced with a mandatory spray program to try to eradicate boll weevils statewide, and perhaps through most of the Cotton Belt," says McMahon.

"I sure want to avoid that, as I would lose my certification for three years, plus a lot of goodwill."

Smarter Marketing

"When we started farming, my husband and I thought people really needed organic products, for health and environmental reasons. With the organic movement building, the time was right," she says.

"We knew we would need a different way of marketing even if we farmed conventionally. There were too many middlemen between the farmer and consumer, and the farmer wasn't getting enough. I knew there was more money to be made for the farmer with some better marketing skills."

Besides running the farm and marketing her own crops, McMahon brokers organic cotton for nine other mid-South growers — and says there is more than enough demand for new growers to join them. Her buyers want 10 million more pounds of organic lint in '93 for a new line of clothing — labeled "O Wear" from Vanity Fair — previewed at a Hollywood fashion show in fall '92.

"I have about 2,000 acres of production lined up for '93. But that's only about a tenth of what we'll need," says McMahon. Her growers include organic farmers trying cotton for the first time, as well as conventional cotton growers experimenting on small acreages. "One of the big barriers to switching to organic production is finding markets. But for cotton, we already have the market — even for growers in transition.

> *"This year, you can run your hand down to the elbow in the 5-year-old beds, and we were dealing with hard gumbo to start with."*

"With cotton, we've found markets for just about everything but the leaves," she says. In '92, she sold 1,300 pounds of bark stripped from the roots. A German pharmaceutical company bought the bark for use in its homeopathic remedies. McMahon will harvest more root bark in '93 if she determines the 50-cent-per-pound price makes it profitable. She also expects to sell moats (cotton lint filtered out during ginning) to the same company. She composts her cotton-gin trash to spread on her certified-organic fields, and plans to bag and sell some next year as an organic soil amendment.

That's not all. She also sells cottonseed to organic growers to plant, and as a protein supplement for organic livestock producers. At her roadside stand, she sells potted cotton plants, cotton-boll wreaths and dried bolls for ornamental arrangements.

Looking ahead to the expanding organic beef market, the McMahons switched from raising hybrid corn to an open-pollinated variety in '92. "We went back to a more nutritious 'OP' variety, as we sell it to organic beef producers in Arkansas," says Lorna. She and her husband are trying to minimize their use of hybrid varieties in all farm operations now. They believe use of hybrids encourages monopolistic seed-pricing practices that are detrimental to family farmers, and note that many seed-supply houses are now owned by chemical companies.

While McMahon has yet to find an organic market for her wheat, she sells most of her soybeans to tofu manufacturers for $10 to $18 per bushel. In '93, some of her beans will be pressed for oil to make organic dyes for the "O Wear" clothing line.

Aside from public receptiveness to chemical-free crops, she attributes her marketing success to making frequent phone calls, growing a variety of crops, having a favorable location, keeping overhead to a minimum and using many marketing outlets.

"We market our high-value products roadside, mail order, direct to restaurants and wholesale. We plan to boost our retail by converting our roadside stand into a grocery store right on the property. We're located on a major highway running through the county, in an area that gets lots of tourist traffic."

She hopes her husband, a contractor, will be able to help full time on the farm starting in '93. The couple expect their high-value enterprises and field crops to generate enough year-round cash flow that they will no longer need John's off-farm income.

"If we can't sell all our crops under our own name, there's also a cannery near here, so we can sell our excess for canning. We also hope to get into schools and nursing homes in the future," says McMahon.

"We're expanding our vegetable and fruit production because there's good money in it. We have 5 acres of raised beds in construction. We'll be automating our drip irrigation there and in our market garden.

"We find the beds don't need that much watering, though, because they're so full of organic matter thanks to our composting program. This year, you can run your hand down to the elbow in the 5-year-old beds, and we were dealing with hard gumbo to start with. I'll only use soaker hose on the perennials soon, and rely on drip tape and black plastic for annuals.

"I'm experimenting with a highboy sprayer to apply compost tea as a foliar fertilizer once or twice to each bed. That might let us stop applying composted poultry manure to the existing beds soon. I'm also experimenting with sea kelp and fish emulsion."

Adding livestock would also allow the couple to stop buying poultry manure, McMahon figures. But she and her husband aren't interested in managing livestock.

"We might be able to talk a neighbor into managing a feeder-calf operation for us. We could put high-tensile wire around the perimeter of our strip-cropping parcel. That would allow intensive rotational grazing of the 2-acre green-manure strips from early spring to winter. Free-range chickens and portable coops could follow the cattle in the strip rotation. Cattle would suppress weeds, then chickens would pick in the manure and get insects and small weeds.

"We're also starting to sell field-grown and container-grown nursery seedlings. We figure if one of our crops ever fails, we'll have plenty of others to take up the slack."

McMahon's goal is to put the farm on a solid financial footing. "We live in a mobile home and drive two 10-year-old vehicles, but we live comfortably. Our three-year-old, Hazen, will enjoy a lot of benefits growing up on a farm. My other two children [one in college, one in the Navy] plan on farming too. They'll be the sixth generation of farmers on this land, and I'd like that to continue."

Bedding plants and ornamentals in 10,000 square feet of greenhouse space, plus 14 acres of high-value crops, provide the McMahons with year-round cash flow.

For the McMahons, using organic methods is more than just a way for their family to continue its farming tradition. They're also committed to spreading the word so that other farmers — and even the whole region — can reap the economic and environmental benefits of regenerative agriculture.

"We're hoping to set an example. By working with other farmers, sharing equipment and developing marketing skills and markets, we're showing farmers who are facing land payments and off-farm employment that there *are* other ways to make more money. We also can help other growers find organic markets for crops they can rotate with cotton.

"When people don't understand something or lack control of facts and figures and just rely on chemical recommendations, it doesn't leave farmers room to make their own judgments. Fifty years ago, farmers were rotating crops without chemicals in our county. The chemical era put people out of work around here. There used to be 30,000 people in this county; now there are just 8,000.

"I don't like the idea of chemical companies taking land away from American families and putting it in the hands of mega-farm owners who drink coffee at the cafe all day. I'd like to see more small-scale farmers get back into making a good living on a few hundred acres."

Editor's Note: *For information about growing cotton for the organic market, contact: Lorna McMahon, Silvertop Farms, R.R. 1 Box 345, Tiptonville TN 38079, (901) 253-8028 or (901) 253-8761.*

Goodbye Corn, Hello Profits

Forage crops and rotational grazing keep this dairy in the black.

GREENWOOD, Wis. — It's getting harder and harder to find corn on Karl and Donna Stieglitz's 70-cow dairy farm — even in the feed ration.

The couple stopped growing corn in '86. It requires too many inputs and carries too many risks, says Karl. "I can buy corn cheaper than I can raise it. Purchasing corn is the easiest, least risky way to eliminate chemicals and avoid playing in the mud here in spring and fall. Forage is the real money crop for dairy farmers, not corn."

Stieglitz has relied mostly on small grains and high-quality hay and haylage from legume/grass fields since the early '80s. Rotational-grazing studies he has conducted since '88 have convinced him to convert many of his hayfields to pasture.

By grazing his herd more, he has cut his standard ration of corn and small grains by more than half. "I've reduced my out-of-pocket feed purchases from $4 per hundredweight of milk to $2.43 per hundredweight. I'm spending less per hundredweight of milk today than when I was trying to grow corn. And I'm feeding less corn. My low-input strategies are really working."

Karl and Donna Stieglitz
Greenwood, Wis.

Started cutting inputs: 1976
20 acres forage sorghum
25 acres forage soybeans
30 acres seed oats
20 acres rye for bedding and feed
95 of 200 tillable acres
318 hayfields and pasture
413 acres + renting 53 acres

Livestock: 70-cow dairy herd
Growing Season: 115 frost-free days
32 inches of precipitation

Stieglitz had plenty of experience with confinement dairying before switching to pasture. He worked summers on his uncle's dairy farm in central Wisconsin from the mid-'60s to the early '70s. After graduating from the University of California-Davis in '73 with a degree in dairy science, Stieglitz started dairying full time on his uncle's farm.

"We farmed conventionally then, and that meant a good share of the farm was planted to corn with all the inputs. I even put attachments on the corn planter for applying rootworm insecticides," he recalls.

Credit Manure And Legumes

Stieglitz wasn't happy buying inputs for row crops, though. In '76, he and his brother installed a 1-million-gallon manure pit to make better use of farm-produced fertilizer. "Dairy farmers shouldn't have to purchase much nitrogen if they take into account their manure and legume credits. Adding legumes to the forages provides N, while manure provides P and K plus additional nitrogen," Stieglitz points out.

In the late '70s and early '80s, however, he was still using about 200 pounds of starter fertilizer (12-24-24 or 9-23-30) per acre for corn, and 200 to 300 pounds of 0-0-60 per acre on sod. He was applying liquid manure at 10,000 to 12,000 gallons per acre for corn and topdressing some hayfields if manure was available.

Instead of concentrating manure on corn acreage to ensure adequate N, Stieglitz figured he could apply manure at lower rates to more of his forage acres by reducing or eliminating corn acreage. "We could apply 6,000 gallons of liquid manure per acre where we would be planting small grains the next spring, and topdress 3,000 gallons of slurry per acre on hay ground."

It wasn't until 1980, when Stieglitz became sole owner and manager of the farm, that he seriously considered giving up corn. He grew 150 acres of corn that year, then began reducing corn acreage in succeeding years. Aside from lowering fertilizer and pesticide costs, Stieglitz also wanted to avoid cutting corn silage in Wisconsin's wet autumns. "On these poorly drained soils, we can really tear up the ground," he says.

"Almost nobody locally was doing anything to reduce inputs at the time. To get more information about cutting chemicals, it made sense to associate with farmers who *were* cutting inputs."

He joined Rodale Institute's Midwest farm network in '86 and dropped corn entirely from his rotation that year. "Since then, I've concentrated more on a forage-based system rather than a grain-based one," he says.

Hired hand Mark Aden says he hasn't had to shovel as much silage since the Stieglitz dairy herd began rotational grazing.

Besides cutting his purchased-fertilizer bill, Stieglitz eliminated his herbicide use as soon as he stopped growing corn. "With no row crops in the rotation, the need for chemicals was gone. All of the acres were either in a forage crop, or else a small grain underseeded with a forage. There was no need for a herbicide.

"I never liked working with chemicals," he adds. "And with two children around the farm, I felt especially good to be rid of them.

"I grow everything solid-seeded now. I rely on plant populations and canopy coverage to control weeds. I work the ground when it's really right, so crops can outcompete weeds."

Grasses Help

Stieglitz includes grasses in his new hay seedings. The grass crowds out broadleaf weeds, and helps hay dry quicker in the windrow, he observes. He seeds 5 pounds of brome or timothy per acre, along with 12 to 14 pounds of a mix of three parts alfalfa and one part red clover. He had been experimenting with planting 2 bushels of barley per acre as a nurse crop, but he found that 3 bushels of oats gives better results. He harvests the oats for grain.

"For the grass, brome has been my first choice. But with all the set-aside and CRP acres nationwide, the price became too high to use brome all the time. I sometimes use timothy instead."

Wet weather is another reason Stieglitz avoids clear seeding his alfalfa. "Clay subsoil causes poor drainage in our county, so two years out of five you'd see root rot and wet heaving damage in pure alfalfa. I include environmentally adapted grass species in my hayfields rather than have to work for a crop. People spend too much money around here trying to get 5-ton alfalfa using lime and potassium, and also see problems associated with atrazine carryover."

Forages Add Flexibility

After three to five years, Stieglitz rotates hayfields to small grains (usually rye) or another forage crop. "I like growing rye for a bedding source. If drought looks likely, it gives me extra feed in spring *or* I can use it for a plowdown," he says.

Stieglitz also likes planting forage sorghum. Drilled in 7-inch rows, the sorghum canopies fast to shade out weeds, so it doesn't require herbicides. It lets him maintain his corn base and also adds flexibility to his field management. "I can plant the sorghum anytime before July. That allows me to take a first cutting of the preceding hay crop (if I think it's good enough) before plowing the stand. Otherwise, I can plant sorghum first thing in the spring."

He uses conventional forage-harvesting equipment to cut, chop and ensile the sorghum. It yields about 15 tons of silage per acre or an average of 5 tons of dry matter per acre. Since he harvests the crop within seven weeks of planting, he avoids wet autumn weather that used to hamper his corn-silage harvest.

Sorghum is a big nitrogen user, so Stieglitz adds plenty of manure and counts on carryover N to help the soybean crop the next year. His forage soybeans average 4 tons of dry matter per acre and 21 percent crude protein.

Stieglitz says the sorghum and soybeans improve his rotation. "As alternative forage sources, they also minimize my risk. I always like having something I can put in the silo."

For part of his dairy feed, Stieglitz buys 3,000 bushels of corn per year. "I feed a half-supplement of grain to the cows when I have them rotational grazing. The supplement is half small grain and half corn (cob corn and dry shell corn), with no protein supplement.

"Corn is always available for sale. In dry years, I still have plenty of forage, where I wouldn't have had the corn. So I minimize risk and use surplus-hay dollars to buy any corn I need."

That kind of thinking proved right on the money

A portable watering tank saves time and energy — for Stieglitz and his herd.

in a dry '88. "I helped the ASCS office appraise corn fields for disaster assistance in fall '88. Most fields had 20- to 30-bushel corn yields and 2-tons-per-acre silage — if the farmer was lucky. I still had plenty of feed.

"In a given year, my purchased corn might cost the same as growing my own would, but I'd just rather not worry about \$20- to \$30-per-acre herbicide costs, \$70 per hour in combine costs and other related expenses."

If reducing your corn acreage, consider the USDA's Integrated Farm Management (IFM) Program Option, Stieglitz suggests. "I've found that by signing up with the IFM option, I am able to hold onto what corn base I still have. The IFM helps me farm the way I want to farm — with a forage-based system. I can be all forage and still have my corn base, plus have 85 percent of the deficiency payments I would have.

"If I ever want to sell the farm, that corn base is one thing people look at. It's a factor in land values."

You can also grow soybeans for forage as part of the IFM, Stieglitz points out. "By growing full-season soybeans, I can harvest some outstanding forage after September."

Grazing Cuts Inputs

Stieglitz began a three-year study of rotational grazing of heifers in '88, supported by funds from the Wisconsin Department of Agriculture's Sustainable Agriculture Program and from Rodale Institute.

"We'd been confinement feeding the entire herd up till then. I relied on haylage, limited hay and a fairly standard feed-mill-balanced ration of grains and supplements based on level of production. The herd only went outside for exercise," says Stieglitz.

He enclosed a 23-acre mixed legume/grass hayfield with high-tensile fencing for the grazing study. The field included mainly alfalfa, red clover, white clover, brome and quackgrass. He subdivided the field into 1-acre paddocks using temporary polywire fencing and fiberglass posts. Fencing cost about \$25 per acre.

For the six-month grazing period starting in late April, heifers rotated through the paddocks every two to three days at a stocking rate ranging from 22 to 31 animals per paddock. The stock were in two groups: small heifers (at least 6 months of age, weighing up to 500 pounds) and larger heifers plus a few dry cows.

"We weighed animals in spring and fall to generate rate-of-gain data for the project," says Stieglitz, who added a few paddocks during summer in an adjacent hayfield. The heifers received no supplemental feed during the grazing season.

Impressive first-year results convinced Stieglitz of the value of rotational grazing. "At 1.94 pounds of average daily gain per animal, heifers showed better weight gains during the pasture months than they had in confinement. We knew we had nice-looking animals even without weighing them.

"With the drought we had that year, those results came as a pleasant surprise," adds Stieglitz, whose rainfall was more than 50 percent below normal in '88. "Even with average rainfall, I wouldn't expect a higher rate of gain, just more forage to store or sell. Rotational grazing allowed me to go into winter with sufficient hay supplies. Without it, I would have had to feed so much through the summer that I'd have had to buy hay for the winter."

"I've reduced my out-of-pocket feed purchases from $4 per hundredweight of milk to $2.43 per hundredweight. My low-input strategies are really working."

He also liked the decreased labor from rotational grazing, compared with confinement feeding. "When you consider all the time spent fixing machinery, preparing silos, waiting for feed carts to fill up and removing manure, the time spent moving fence was actually shorter."

Learning To Pasture

Stieglitz had a few problems at first with animals escaping the temporary fencing, but figures his own inexperience was mostly to blame. "I was unfamiliar with the setup, never having done pasturing before. I was trained along with the animals!"

He eventually replaced the thin electric temporary wire with a wide, orange electric tape. "The improved visibility of the tape solved most of the problem. I suggest using the plastic tape with younger stock — that, and doing a proper job of orienting them to the fence," says Stieglitz.

Drought again slowed pasture growth in '89, and heifers received supplemental feed equal to one-third their normal ration. Animal weight gains were less than the previous year, averaging 1.65 pounds per day during the grazing period.

In '90, Stieglitz added a mobile watering system so cattle needn't return to the barn for water. For less than $10 per acre, he laid black PVC pipe aboveground and used a light, rubber watering tank that could be positioned between two paddocks. He moved the tank frequently, avoiding pasture damage that heavy foot traffic can cause around permanent watering stations.

Forage protein levels consistently exceeded 20 percent for samples taken during summer months in '90, compared with 15 percent in '88. The stock received no supplemental ration in '90. "By the third year of grazing, I actually had *too* much quality when I turned in those young stock. I saw less rate of gain, but a wet year also reduced grazing activity in '90. Animal weight gains averaged 1.28 pounds per day that year," says Stieglitz.

Over the course of the three-year study, animal weight gains averaged 1.64 pounds per day. "That's pretty good performance and with minimal or no supplements," reports Stieglitz. "I obtained better forage from pasture than I was able to put up mechanically harvesting hayfields." He also saved a total of about $4,000 per year on feed, fuel and equipment. He didn't track labor savings, figuring that was an internal resource.

With a better handle on pasture quality, Stieglitz couldn't wait to start his milkers on rotational grazing. "After three years of running heifers on pasture, I figured my cows deserved to be out there for five to six months of the year getting fed on the hoof rather than with a tractor and silos," he says.

Cows Deserve Grass

In '91, he fenced 79 additional acres of legume/grass hayfields, using single-strand high-tensile wire to establish long, narrow, semi-permanent paddocks. "You reach a point where you don't want to move fence," says Stieglitz, who had been spending as much as two hours a day moving a lot of portable fencing. Now, he makes 1-acre subdivisions by moving just 150 to 250 feet of temporary fencing when pasture conditions and labor permit. That makes it easier for Stieglitz to rotate paddocks daily in dry summer months when pasture growth slows.

To eliminate muddy lanes that led to tedious udder washing every milking, Stieglitz had a clay lane built to the pasture area. "The cows were traipsing through so much mud in '91, I was tempted to stop grazing them. Properly planning your cow traffic is one of the most important considerations in pasture planning," he says.

Stieglitz also recommends using a portable watering tank that's large enough to meet the herd's needs. "Most grazing consultants say watering becomes less of a social function when the water supply is fairly close. But I still have social cows — most of the herd goes for water. I hope that decreases over time."

To ensure adequate milk production during the grazing study in '91, Stieglitz fed cows a daily half-ration of protein supplement. With the better-quality forage, he also reduced his feeding of corn and oats by 20 percent, with each cow receiving about 8 pounds of dried corn (half cob corn, half shell corn) and 6 pounds of oats. "I cut about $5,000 in dairy-cow feed costs last year, with my biggest savings from reduced supplementation," he says.

Stieglitz also saved about $5,000 in labor, fuel and equipment, his figures show. Reduced electricity use alone (for lighting, ventilation, and silage loading and unloading) amounted to $1,200 during the six-month grazing period.

Rotational grazing really helps you cut your milk costs, says a happy Stieglitz.

"My rolling herd average stayed the same, at about 16,500 pounds. I was cautiously optimistic it might go up, as some dairy farmers have reported in their first year of rotational grazing," he says.

Figures Look Good

Stieglitz hasn't had any worries about poor-quality feed. "Pasture forage samples averaged 24.9 percent protein and a relative feed value of 150. So I know I'll get good milk production as long as I manage my grass properly and have proper forage utilization," he notes.

Cows maintained a good body condition during the study. Average ratings ranged from 2.97 to 3.08, with no significant differences seen during the grazing season.

"I can't say if grazing improved the breeding, as breeding has never been a problem for my naturally bred heifers and AI-bred cows. I'll see occasional conception problems in the milking herd, but nothing of concern," observes Stieglitz.

"My vet bill is nominally less today than in '88. I don't push my cows hard. That reduces their stress level and leads to lower vet bills, which consist mostly of monthly herd-health checks and vaccinations. There's very little other health work involved. Now that my cows are on pasture, I think there'll be even less vet work, as they're getting fresh forage and natural vitamins.

"My ultimate goal is to get my ration to include forage and whatever small grain I have on hand. I'm not there yet, but that's where I'm heading. I'm already down to feeding just 8 pounds of grain per day and no protein supplements when the herd is grazing.

"I see the day when I won't need the 53 acres I'm renting. My debt load is coming down each year, and that will improve my profitability."

Stieglitz credits rotational grazing with helping him and his family weather low milk prices in '91. "Saving $15,000 by having the entire herd except baby calves on rotational grazing certainly didn't hurt. And that's a conservative estimate. We didn't count all the fertilizer savings we could have," notes Stieglitz.

"In '91, we delayed some equipment replacements, but we also need less equipment than before. And now that we've gone to grazing, we're putting about two-thirds of the previous use on our equipment."

Carving A Path

Stieglitz will continue to document production costs for rotational grazing in '92 and '93, thanks partly to additional research funding from the

Wisconsin Department of Agriculture's Sustainable Agriculture Program and Rodale Institute. He believes that on-farm research such as his, along with farm field days, winter workshops and Congressional lobbying, are needed to promote sustainable farming methods nationwide.

"It costs money for farmers to go to Washington, but sometimes you have to do it to get your message across," says Stieglitz, who testified before Congress in January '92 about the need for funding to train Extension agents in sustainable methods.

He also is active in a number of civic and farming organizations at the state and community level. He is the county supervisor for Clark County, Wis., and a member of the citizens advisory council for the University of Wisconsin's (UW) Center for Integrated Agriculture Systems. He's also on the technical steering committee for UW's Agriculture Technology and Family Farm Institute, which is helping to improve and maintain the quality of rural communities.

"Too many farmers are told that maximum production is the only way to go to survive in farming. I've been discovering more and more that what works with the least amount of intervention with nature becomes the most profitable strategy for me.

"You don't think you do something just because others do. But as you take a different approach you can't help but question whether what you're trying to do will work in the long run. Well, it's working for me.

"Without row crops, I've pretty much eliminated soil erosion and compaction. Now, I'm not only below 'T,' but I'm actually building and improving soil.

"I'll let others produce the corn and all that goes with it. We'll continue with what produces the money for us: quality forage."

To keep cows out of a "sea of mud," Stieglitz built a clay pasture lane from the barn to the main field lane, saving about 45 minutes in cleaning time each milking.

Organic Farming: An Investment Or A Gamble?

By fine-tuning inputs, labor and machinery, these large-scale farmers hope to find out.

WABASH, Ind. — To increase the time they devote to organic farming, ridge-tiller Barry Eppley and his father, Dean, made a tough management decision in '91. They stopped banding herbicide on half of their 1,800 acres of corn and soybeans, so they could eliminate one and possibly two cultivations. They went back to broadcasting chemicals for weed control.

"We had too much weed pressure when all the rows were banded," says Barry. "Because of wet springs in '89 and '90, we found it tough to schedule all the cultivation. When we were in the fields, we often had two cultivators going at the same time, which put a tremendous amount of pressure on us."

Barry and Dean Eppley
Wabash, Ind.

Started cutting inputs: 1983
900 acres corn
900 acres soybeans
60 acres wheat
140 organic acres:
20 acres winter wheat
60 acres soybeans
10 acres corn
50 acres set-aside

2,000 acres

Growing Season: 160 frost-free days
36 inches of precipitation

Since '83, the father/son team has been using ridge tillage on their 2,000-acre cash-grain farm in northeast Indiana. They also have worked hard to reduce their synthetic inputs.

"We stopped using insecticides in the mid-'80s by eliminating corn after corn. We were willing to lose some corn base to get to a corn-soybean rotation. And we reduced herbicide use significantly by banding a 10-inch width instead of broadcasting on our 30-inch rows," Barry says.

They also cut their fertilizer rates. "We still broadcast some P and K to maintain a 'medium' range for our soils. But we have some acres that haven't received *any* P or K for seven years, without a significant decline in soil-test levels," notes Barry.

"Depending on soil type, we had been applying anywhere from 200 to 250 pounds of synthetic nitrogen per acre in the early '80s. We cut that to 85 to 115 pounds per acre," adds Barry. The move saved them about $25 per acre.

"That's a little on the low side of what most people would recommend for nitrogen on corn after beans," Dean points out. "But three years of late-spring soil nitrate tests starting in '89 did not show any yield response for us beyond applying 45 pounds of nitrogen per acre." Weather has been a factor, and adding just 45 pounds of N per acre wouldn't be a reasonable long-term rate without additional N from legume covers or animal manures, Barry says.

Organic Premiums Appealing

Looking for additional ways to reduce inputs, labor and machinery costs, the Eppleys met some farmers who were successfully raising row crops organically on hundreds of acres. "Through contacts in the Indiana Sustainable Agriculture Association and the Organic Crop Improvement Association (OCIA), we learned more about organic practices and buyers. The crop premiums looked good. So we decided to try some organic crops in '89," says Barry.

"Essentially, we were motivated by economics. If we could purchase fewer inputs and still maintain productivity, we would be better off, economically *and* environmentally."

They started by plowing down a 44-acre field that had been planted in alfalfa/grass set-aside in

'85. It had been chemical-free since then, so it was easy to have OCIA certify the field. "The acres weren't set aside with the intention of certifying them. Had that been our intent, we would have tried to increase the alfalfa and reduce the grass," Barry says.

"We planted 44 acres of adzuki beans in '89 and thought we had a crop failure when we got just 4 to 6 bushels per acre. Then we found we could sell them for $1 a pound rather than the 44 cents a pound we initially expected. So with a strong market and organic premium, we wound up doing quite nicely."

Pleased with their initial return, the Eppleys increased their organic enterprise to 90 acres by having additional set-aside acres certified in 1990.

"That year, we had 60 acres of VINTON 81 clear-hilum tofu soybeans and about 30 acres of adzukis. Wet weather delayed both plantings and weed pressure hurt the adzukis, resulting in small plant sizes and an even lower yield than before. We didn't harvest many of them," says Barry.

Weeds were more manageable in the VINTON fields. For all their organic acres, the Eppleys were relying on spring tillage using just a chisel plow and disk to maintain as much residue cover as possible until planting. "That's how we chose to address weed control initially," says Barry. "We used existing equipment for tillage and cultivation. We rotary hoed two or three times, followed by two or three cultivations. When we've had the opportunity, we've had reasonable weed control with mechanical cultivation."

The VINTON beans yielded 27 to 30 bushels per acre. "That really got us going, because they performed and yielded more like what we're accustomed to seeing from soybeans. And we sold them for about $11 a bushel."

Although market demand was strong, the Eppleys chose not to contract with buyers until after harvest. They relied on word-of-mouth networking with other growers, with OCIA contacts and with wholesalers. "We found it easier to plant and grow, then clean and bag *before* trying to sell, rather than establishing and contracting a price in advance. It was a gamble, but so was our whole organic program," explains Barry.

More Weather Woes

Encouraged by their '90 results, they diversified with five different crops on their 90 organic acres in '91. They grew about 20 acres of blue corn, 20 acres of black soybeans, 10 of winter wheat, 10 of spring wheat and 30 of buckwheat. It was their first attempt at growing blue corn and black soybeans.

A special strain of fast-growing grafted black walnut trees provides Barry Eppley and his son some diversification into a high-value crop. Corrugated plastic tree shelters afford protection and branch training to the 140 trees Barry planted on an acre behind their home. Promoters say the Purdue No. 1 trees can be harvested for veneer after 25 to 30 years — about half the time of native black walnut.

Weather proved problematic again, with May rain causing poor germination of buckwheat following adzukis. It was too wet to rotary hoe most of the other crops. "By the time we were in the fields cultivating, there were too many weeds growing. We couldn't get equipment into the black soybeans, for example, until what would have been time for second cultivation. That same field gave us very clean VINTON beans in '90," says Barry.

They managed to harvest 20 bushels of black soybeans per acre, but just 6 to 7 bushels of blue corn per acre. "We only bothered to harvest about 15 of the blue corn acres, and that wasn't until February '92. We should have harvested in October, when we might have gotten 25 bushels per acre. But we had some rain then and the fields needed to be drier before we could harvest. It was still wet in December, and by then we already had losses due to wildlife," Barry recalls.

Their organic wheat fared better in the field than at the market. "The organic hard red winter wheat was the best wheat we raised that year, yielding 44 bushels per acre. '91 was a tough wheat year around here, averaging 30 to 35 bushels per acre instead of the normal 65 to 80."

They delivered that wheat for $6 per bushel, but wound up waiting more than eight months for payment. That surprised them, as it was the first time they had a payment delay in the organic market. The buyer was one they were happy to deal with, who had always paid promptly in the past.

Their organic spring wheat yielded close to 30 bushels per acre, but they had trouble selling it. Due to a cool, wet, cloudy May, protein levels weren't as high as the organic market wanted.

In fall '91, they planted 50 acres of spelt (a winter-wheat relative used in health-food pastas) and 20 acres of winter wheat. They hoped to have better success in '92 with some of the crops they had trouble with the previous year.

In early '92, however, they saw that the spelt wasn't doing well. "A warm spell in November followed by a sudden deep freeze in December had apparently weakened the spelt and encouraged stalk browning. By spring, it was obvious the grain wasn't worth saving," says Barry. The Eppleys made that field set-aside acreage, and left the spelt as cover, planning to mow it off eventually. The field remained organically certified.

The 20 acres of winter wheat came through the winter better than the spelt. Rather than plant blue corn or black soybeans on their 70 other organic acres, however, the Eppleys planted 60 acres of VINTON 81 soybeans and 10 acres of dent corn. They knew what they could expect with the beans. They wanted to see how well solid-seeded corn could outcompete weeds.

"We held the organic production to 90 acres in '92. We couldn't let our organic experiments be a major financial drain, and '91 clipped some of our wings. We didn't see an overall return on the organic acres in '91 the way we did in '90," says Barry.

"We're convinced that organic production is possible and offers a potentially good return on investment. But we haven't reduced the risks enough yet to conclusively justify larger acreages. Other farmers are doing it, but we're not there yet."

Cover Crops A Priority

The Eppleys are working to hone their organic farming practices, however. "We want to gain additional experience in crop nutrition and controlling weeds without chemicals," says Barry. "Without any livestock, our options are limited. Cover crops are about the only source of nutrition we can control on the organic acres, so we'll have to address that issue better.

"We've had some trouble broadcasting vetch into soybeans. We're doing more test strips, and frost seeded vetch into the organic wheat acres in spring '92. We might consider direct seeding alfalfa on some acres, which we've done before.

"Aside from good weather to allow timely cultivation, cover cropping would be the most realistic way to reduce weed pressure on our organic acres. We don't like feeling helpless when the weather doesn't cooperate. Organic certification

Photo Courtesy of Ridge Till Hotline

Barry and Dean Eppley like to see mellow soil in their fields, whether it's from zone tilling ridges or from a diverse rotation using cover crops and organic methods.

Photo Courtesy of Ridge Till Hotline

With their modified 16-row Buffalo planter, the Eppleys zone-till plant corn and soybeans on 1,800 acres of ridges.

has kept us from reaching into the quiver and pulling out a herbicide.

"We also want a rotation that includes small grains or even a fallow season of cover crops as a way of breaking weed cycles. Sometimes it doesn't look like we're planning far enough in advance on the organic rotation. But we've got a year of wheat behind us, so that part of the rotation is established. We would be more comfortable with organic beans and wheat rather than a year of corn. Insects haven't been a problem — weeds are our only pests."

Ridge-Till Pioneers

The Eppleys are well-known in Indiana for their innovative approaches to ridge tilling large acreages. In '91, they modified their 16-row Buffalo ridge-till planter to allow zone-tilled planting into their ridges. They replaced the ridge-till cleaning units with three Rawson fluted coulters per row, mounted on the front toolbar, well ahead of opener assemblies.

Together, the three coulters work a strip of soil approximately 10 inches wide. A 1-inch-wide coulter slices up corn stalks and root clumps directly on the old row top. Offset 3 inches to each side of the first coulter, two trailing 2-inch coulters stir the soil without pushing it into row middles.

The Eppleys like the way the coulters make it easier to till weather-hardened ridges. They think zone tillage saves more soil than ridge tillage, since they're only tilling the tops of ridges that are rebuilt every two years.

"We wanted to plant into mellow soil on top of the ridge, rather than skimming it into the furrow," explains Dean. Although the coulters leave some residue on ridge tops, plugging of planter assemblies hasn't been a problem, adds Barry. The Eppleys made sure of that, by mounting new International Cyclo opener assemblies on a second toolbar 12 inches behind the first.

Photo Courtesy of Ridge Till Hotline

Although Rawson coulters can increase weed pressure, they help keep fluffy soil on the ridge tops instead of pushing it into row middles.

At $1,400 per row for the coulters and planter assemblies, the modifications cost $22,400 for their 16-row planter. But not having to make new ridges each year, plus fewer cultivations, will allow for a quick payback, the Eppleys figure.

"We still like to cultivate at least once for the aeration benefit, if we have the opportunity. The cultivation leaves a 3- to 4-inch hump, adequate to help elevate the rows for ridge planting next spring," says Barry.

"Because we broadcast some herbicide again, however, we didn't cultivate 600 of the soybean acres at all in '91. And we've eliminated two cultivations on up to 300 corn acres that don't need any sidedress N. In '92, these fields got brand new ridges, a break in herbicides and the aeration benefit of cultivation.

"On our better soils, we sidedress only when late-spring soil nitrate tests indicate a need for additional nitrogen. So we're eliminating trips between planting and harvest on lots of zone-tilled corn ground, especially in the year we're not ridging.

"Ridge-tillers may not appreciate what we've done with our ridges. We've eliminated the ability to peel off the tops of ridges, which is one way of tilling the ridges."

By loosening soil and stimulating weed seeds just before planting, Rawson coulters can increase weed pressure, Barry points out. "The coulters also make it a little more difficult to incorporate herbicides, because if applied immediately ahead of planting, herbicide is still on residue and may be thrown out of the row zone by the coulters.

"The Rawson coulters are not consistent with strict ridge-till or organic systems. Although they're good for *reducing* herbicides, the Rawsons aren't intended to eliminate herbicides. We haven't eliminated herbicides on our ridge-till acres, although we have reduced a good deal of our herbicide use via banding on many of our acres. We're also trying to include wheat in a three-year rotation with corn and soybeans on more of our lighter soils.

"All told, we're cutting inputs on many acres without sacrificing yield. Plus the time saved allows us more time with our families and more time to manage our organic crops."

Exploring Other Options

The Eppleys are experimenting with using paper-fiber waste as a source of organic matter. They're working with a local paper mill that provides free delivery of an organically acceptable paste made from newsprint paper fibers too short to recycle.

"We've tried some light applications on about 150 acres of ridges. Although it clumps and globs a bit, the paste doesn't need additional tillage and the earthworms like it. It requires some field nitrogen to decompose, though, so we can't grow corn or wheat after applying it. We try to follow with soybeans, which do well with it. We think it will help soil organic-matter levels in the long run and we could eventually use it in a composting operation," says Barry.

To help cut herbicides further, especially in soybeans, the Eppleys acquired a new Hiniker Econ-O-Till Seeder in March '92. "It's very cost-competitive with no-till drills. We'd like to look at planting more of our acres as modified no till, and solid seeded about 400 acres of beans in '92 with the Hiniker," says Barry.

The Econ-O-Till combines Hiniker's Model 5000 high-residue cultivator with a detachable, pull-behind, air-delivery planter unit that provides randomized seed placement behind a wide sweep. Hiniker staff say the weeder-seeder works well in a wide range of soil types. You can use it on heavy loam or sandy soils, and in wet and dry conditions. (Contact: Hiniker Co., P.O. Box 3407, Mankato MN 56002-3407, (507) 625-6621.)

"Following corn, we'll also seed a rye cover on 50 to 100 acres in fall '92, including some replicated plots," says Barry. "Early the following spring, we'll cut the rye off with the Hiniker and lay the soybeans under the decaying mulch. The rye would provide an allelopathic mulch and the randomly planted beans should canopy quickly. There's an opportunity to cut herbicides in soybeans this way.

"On our soils, ridge tilling still has a place, especially in wet weather when it takes time for soil to dry out," notes Dean. Like his son, however, he's intrigued by the potential of tillage modifications and organic farming practices to reduce the use of expensive inputs. In early '92, the USDA appointed Dean to one of 15 positions on the National Organic Standards Board. The board will play a pivotal role in defining federal standards in the increasingly important organic farming industry.

As part of Rodale Institute's Midwest Farmer Network and the Indiana Sustainable Agriculture Association, Barry Eppley is also playing an important advocacy role in farming. He often shares information about sustainable practices, input reduction, growing and marketing organic crops, and fertilizer comparisons performed on the Eppley farm. A Purdue University farm tour that featured the Eppley farm in '91 attracted more than 350 participants.

Will the Eppleys expand their organic acreage? Perhaps. "The organic industry isn't developed enough yet," Barry notes. "A lot of economic factors are at work.

"Tools like the Econ-O-Till will allow us to continue to reduce inputs. If we perfect our organic practices, we'll be working in the direction of more certified acres as long as commodity prices encourage us to do so."

'Economics Is The Key'

Rotations and cover crops help this Extension agent reduce farm-chemical use.

DOVER, Del. — Mark Davis wants to show other farmers there really are profitable alternatives to many conventional cash-grain farming practices.

In '91, Davis began working full time off the farm, as an ag business management specialist for the Delaware Cooperative Extension Service of Delaware State College. He cut his own acreage in half. He rented out 205 acres, while continuing what he had started three years earlier on 195 other acres of no-till corn, soybeans and wheat: banding herbicides, improving his rotation and introducing cover crops.

"I was reading *The New Farm* and wanted to try some different farming practices. The farm had been operated conventionally for years before I took over in '81, and crop rotations had gone by the wayside. I'm trying to get them back, here and elsewhere in our area," says Davis.

Mark Davis
Dover, Del.

Started cutting inputs: 1988
65 acres corn
65 acres soybeans
65 acres wheat
195 acres

Growing Season: 190 frost-free days
45 inches of precipitation

"My goal is to enhance profits while turning the farm into a research-oriented one, working with cover crops, no till and fewer chemicals."

No Yield Loss

In summer '88, Davis began banding herbicides instead of broadcasting at conventional rates. He also bought a Buffalo high-residue cultivator.

"I cut my herbicide bill by two-thirds by banding 12-inch strips. I saved about $40 per acre on 200 acres of double-crop soybeans I planted following wheat. That paid for two-thirds of the new cultivator in one year.

"I knew banding was a proven way of cutting costs and producing a profitable crop. And it gave me the same yields. I haven't seen *any* yield reductions, because I've been implementing alternative practices gradually," says Davis.

In '89, he began a three-year rotation. He plants no-till corn the first year, followed by ryegrass in fall. The next spring he no tills soybeans, then fall plants winter wheat. After wheat harvest in summer and a two-month fallow period, he seeds crimson clover or vetch in early fall. Davis likes the way he rarely needs insecticides with a rotation like this. About the only insect pests he's had a problem with have been grasshoppers, which required some spraying in '91.

"After banding, it makes sense to start improving your rotation and trying cover crops, then including leguminous covers to see if you can cut your fertilizer bill," he says.

"It's like fitting together puzzle pieces. I want to show that if I can farm 195 acres part time, then full-time farmers can operate more sustainably on a larger scale."

Like many part-time farmers, Davis figures he probably can't eliminate chemicals. "But I sure can cut back on them."

Now that he's banding, his herbicide costs are minimal on his small number of acres, Davis points out. "I want a clean row, so I expect to continue banding herbicides. I use label rates for Roundup (at $10 per acre) to kill clover when planting no-till corn, plus a quart of Bicep banded over the row to control grass and broadleaf weeds.

"I still want to produce a good crop. I don't want to see 65 acres of weed-infested corn, but I want to use as little chemical as possible. If I use any chemicals, I want to use them properly.

"I'm no tilling 65 acres of soybeans into ryegrass. No-till beans with covers works best for me. I don't have time to do extensive tillage. And any tillage promotes weeds. The less tillage I do, the fewer weeds I see and the better the banding works. I

Finding success with herbicide banding, cover crops and a three-year rotation, Mark Davis will try growing some no-till soybeans without any herbicides.

spray ryegrass with Roundup to kill it. Then I band 1 quart of Dual with a pound of Gemini over the row to control weeds and grass in the row.

"I have no real weed problems, as I also cultivate corn and beans twice. We had a dry spring in '91, and everybody stopped planting soybeans around here. Those who planted early had weed problems. Nobody's chemicals worked on full-season soybeans. At 34 bushels per acre, my no-till beans worked well, as did some soybean test plots using just Roundup. In '92, I'll be trying a few acres of corn and soybeans without *any* chemicals."

Covers Supply N

Davis doesn't fertilize soybeans at all following corn and ryegrass. In most years, full-season soybeans have been yielding a county-average 35 to 40 bushels per acre in his no-till fields.

Before planting no-till corn, Davis burns down crimson clover covers, which provide plenty of residue and boost nitrogen availability. In one 30-acre field where he had drilled the clover after tilling the ground, the cover grew so thick that Davis worried he wouldn't get a satisfactory corn stand. "I bought Acra-plant Trash Whippers, planning to work the clover residue and prepare the seedbed for replanting. But then we got a heavy rain and I decided the yield reduction due to late planting didn't merit replanting," he says.

Biomass cuts taken from the clover in that field averaged more than 6,700 pounds of dry matter per acre. At 3 percent nitrogen, that would be about 200 pounds of N per acre.

Davis doesn't incorporate the clover, so it provides a slower nitrogen release than other tillage regimes might allow. He wants to reduce his use of synthetic nitrogen for corn, but he wants more test data before actually doing that on many acres after a legume.

"Following clover, I still put on the same starter fertilizer I always have — 24 pounds of N per acre from 300 pounds of 8-13-27. In some replicated test plots as part of some Rodale network studies in '92, I'm varying sidedress rates at cultivation to 20, 40 or 60 pounds of N per acre, compared with my usual 100 pounds. I don't want too much N on my corn, so I'm also using the Cardi quick nitrate test when corn is 6 to 12 inches tall.

"I've also started working with vetch. In fall '91, I planted about 30 acres of vetch. It gave a thinner stand than I wanted in spring, possibly because I didn't overlap the spinner passes enough when planting. We also had a dry fall. With a good stand each spring, I wouldn't fertilize corn at all except for starter fertilizer down the row at planting time."

Davis figures it will be two or three more years before he actually cuts his sidedress rate for corn. He wants to see how the variable sidedress rates following clover compare with the use of hairy vetch without any sidedressed N.

"Vetch fixes more nitrogen than clover, so even though vetch seed costs me more than clover does — 75 cents a pound vs. 60 cents a pound — the vetch is more economical. I'm toying with the idea of growing my own vetch seed.

"I've come to the conclusion that the seed cost (at 20 pounds per acre) is enough that you really need to drill your cover crop to ensure a good stand, rather than broadcasting it. I can't justify buying a new no-till drill for 200 acres, but I know I could find a good used one for less than $6,000.

"Cover crops aren't costing me any money if I get a good stand. You might be able to just broadcast Roundup to kill a cover such as clover, and then avoid banding any other herbicide over the row, depending on the weed-controlling mat you get from the cover. If that's the case, your economics definitely work well.

"It's hard to put a dollar figure on the benefit of building organic matter by using cover crops. If you factor that in, then covers always pay for themselves. Even if you're getting a so-so nitrogen contribution, you're still getting organic matter as a bonus."

Test Values Encouraging

Organic matter is certainly important on sandy coastal soils such as Davis has. "Not surprisingly, my organic-matter test values have started creeping up a bit. They're currently averaging about 2 percent. I'm also adding organic matter by growing HOFFMAN 28 winter wheat for its straw. That variety averages more than 100 bales of straw per acre that I've been putting back on the ground," he says.

Davis soil tests all his fields in late fall, something he's done for years. To maintain pH on his sandy soils, he limes every three years. He keeps other inputs to a minimum. "P and K have been stable, and I'm only adding P and K in the starter fertilizer for corn," he says. That shows the cover crops are helping maintain nutrient levels while cutting fertilizer costs, he points out.

"Everyone likes higher yields, but I don't want to trade dollars for inputs to get yields. I want to grow the most for the least."

Crimson clover residue provides organic matter, slow-release nitrogen and weed control, plus it conserves soil moisture.

Davis irrigates his corn, varying the amount of water he adds each year based on the season's rainfall. "Again, I'm not shooting for big yields. In our area, 100-bushel corn is the average dryland yield. Irrigation is an insurance policy that helps me get 125- to 150-bushel corn."

Extension Can Help

The Delaware Cooperative Extension Service has been very innovative, says Davis. "This area is one of the leaders in no till, due largely to Extension encouragement since the mid-'70s. No till works well on our light, sandy, rock-free soils.

"We're constantly checking out new ways of doing things. Most Extension personnel in this area are open-minded when it comes to new ways of farming.

"We're seeing some reluctance by farmers to cut chemicals and cultivate again. Many farmers recall previous cultivation practices, but have gotten away from it. Many need better management skills and improved record keeping before they can effectively get a handle on sustainable methods. But it's a good goal to work toward. They can do it."

▼ *You don't need high sidedress rates if corn follows a good legume cover crop, Delaware Extension agent Mark Davis points out.*

Planning, money management and cover-crop establishment are generally the areas you have to focus on when switching to more sustainable methods, Davis emphasizes. "Rotations, equipment and time management all affect how well cover crops work on specific farms.

"My farm time is spread out better now with a three-year rotation. I don't think I'm working more total hours. I can work from May to November at a steadier pace.

"My income has remained relatively the same with sustainable practices. I haven't been in any federal commodity program for years, either. I don't want to wonder where my money is coming from.

"Economics is the key. If you can show a farmer where he can spend less and grow a good crop with more in his pocket after harvest, that's where we show the benefits of sustainable agriculture."

▶ *By selling corn grain for local milling, farmers can keep transportation costs to a minimum, Davis says.*

'You Have To Be Patient'

On-farm research helps this hog farmer cut costs and control erosion.

NAPOLEON, Ohio — With his sandy soil, weed-promoting hog manure and more than 250 acres of no-till corn and soybeans, Rich Bennett knows a thing or two about farm chemicals.

Since joining the Rodale Institute's Midwest farm network in the mid-'80s, however, Bennett has slashed his herbicide and fertilizer costs by 50 percent. He has honed his use of a rye cover crop to reduce weeds in soybeans and is scaling up his use of hairy vetch as a nitrogen source for corn. By pasturing more of his pigs, he knows he could lower his feed costs even further in the '90s.

"My chemical bill has been cut in half — or more — since I started cutting inputs," says Bennett, who's been farming full time on the family farm since '73. He and his wife, Nancy, have a 50-sow farrow-to-finish operation on their 600-acre farm in northwest Ohio. They feed about 40 percent of the corn harvest to the hogs in a typical year, along with purchased soybean meal.

Rich and Nancy Bennett
Napoleon, Ohio

Started cutting inputs: 1984
250 acres corn
250 acres soybeans
100 acres wheat
600 acres

Livestock: 50-sow farrow-to-finish operation

Growing Season: 155 frost-free days
34 inches of precipitation

The Bennetts farmed conventionally until the mid-'80s. When Rich hired an agronomist to do some soil tests in '84, the results surprised him. "He said I didn't need phosphorus or even much potash. With my sandy soils, I thought he was nuts," says Bennett. "But I was putting so many inputs in, it was eating me up financially. So I had all my fields tested. I wound up cutting P and K starting in '84, mainly in fields where ample hog manure had been applied.

"I was concerned about possible yield loss, so I did some side-by-side comparisons. I'd been programmed to apply potash annually. I was really pleased to see no differences in the side-by-side comparisons. Within two years, I was cutting back on purchased P and K quite a bit."

Bennett also started using plant tissue tests in '84. "It may not be the best test, but it's a good check on whether enough NPK and micronutrients are getting to the plant," he says.

"The consultant understood where I wanted to get to in terms of maximum yields with minimum inputs. He was big on micronutrients, though, and told me I needed more sulfur on one field. I faked him out. After I said I had added some, I asked him to tissue test the field. After getting the results, he said the sulfur level had improved. But when he saw my face he knew I hadn't really added any. He was convinced then too. You need to know your farm and your crop results before making micronutrient recommendations based solely on tissue tests."

In spring '86, Bennett went to his first Rodale Institute 'Take Charge' workshop. "That really opened my eyes and ears. There and in *New Farm* magazine, people said you could cut way back on purchased inputs without losing yield."

On some of his wheat acres in spring '86, Bennett eliminated the customary $11 worth of sidedress N per acre he'd been applying. He was growing wheat as bedding for his hogs rather than as a cash crop, because he figured it wasn't worth spending more on inputs to boost yields above his average then of 45 bushels per acre.

After harvest, he took an 8-inch soil sample, plus a deeper, 16-inch sample to see if eliminating the sidedress had depleted the N levels. "Seems the lower we went, the more nitrogen we picked up. So we had enough for the wheat without the sidedressing," Bennett concluded.

His testing program and use of comparison strips convinced Bennett he wasn't losing as many soil nutrients through crop removal as he had thought. On many of his fields, he was applying 3,000 gallons of liquid hog manure per acre, supplying about 45 pounds each of N, P and K per acre.

"Sandy soil and hog manure really equals weeds, though," Bennett says. He wanted to minimize his use of herbicides, without suffering yield losses. "In '86, I started banding Lasso-atrazine on all the corn, with real good success. I banded Roundup-Lasso on 20 acres of no-till soybeans in '90 and was banding 150 acres of no-till beans by '92," he notes. Whether he's banding or broadcasting herbicide, he also typically cultivates twice.

On-Farm Research Helps

As part of the Rodale Institute Midwest farm network, Bennett started testing input reductions on a small scale using replicated plots. He began comparing different rates of sidedress N in corn in '88, for example. "People saw flags on my test field and would ask about them. They liked the idea of cutting costs and were pretty interested in what I was doing, especially when they heard I was getting comparable yields on the different nitrogen plots," he says. The trials showed that many of the plots benefited from plowing down a mixed stand of red clover and sweetclover before corn.

Bennett was reluctant to cut his sidedress rates for corn, however. "I was a little antsy about our sandy soil's nitrogen needs and wanted a good crop. I was trying to make every step count, though. I started putting sidedress N on corn with the cultivator at last cultivation. That saved a step."

Bennett tested the Iowa State University late-spring soil nitrate test in '90. He saw a 12-bushel yield loss in a corn demonstration plot when he applied no N at the recommended sidedress-N cutoff level of 21 parts per million (ppm). "Back then, they were suggesting 21 ppm as the sidedress cutoff, but we learned later that the 21-ppm cutoff didn't really apply to sandy soil. On-farm results, including ours, indicate the test probably works better at 25 or 26 ppm.

"Fortunately, [Midwest network coordinator Dr.] Jim Tjepkema didn't want us to go full bore with sidedress reductions until we had more experience with the spring nitrate test. I'm glad he made that recommendation," says Bennett.

"Using the late-spring soil nitrate test at the higher cutoff rate, I've seen comparable yields in demonstration strips with and without sidedress in '91." On those plots, Bennett compared fall-injected liquid manure, winter surface-applied liquid manure and a hairy vetch plowdown, with or without 60 pounds of 28-percent liquid N sidedress per acre.

Bennett prefers to use 28-percent liquid N when sidedressing, having had a bad experience with anhydrous ammonia in '79. "The hose broke and just went crazy," he recalls. "Fortunately I had a cab tractor and could get upwind. But that put the scare into me and I went strictly with 28-percent for purchased N until recently. In '91, I used anhydrous in a comparison check and found that it showed some promise on sandy soils low in organic matter. So we're testing anhydrous again in '92."

In manured cornfields and those benefiting from legume N in '92, Bennett planned to base the side-dress rate on the late-spring nitrate test. "Any field testing at or above 25 ppm won't get any additional N," he says.

No-Till Beans With Rye

Erosion potential has been a concern of Bennett's. "I always had a problem with wind erosion. My dad used a lot of rye to hold the sandy soil. I started thinking about some of the ways old-timers worked and started to see how alternative methods were benefiting the soil. They let you maintain organic matter and conserve moisture," he says.

Bennett decided to minimize soil disturbance by doing less fall tillage. He also began to experiment with cover crops. He stopped chopping corn stalks and chisel plowing much of his corn stubble in '89, when he started no tilling some of his acres. "I started broadcasting rye with a fertilizer spreader, then disking in the rye and planting no-till soybeans in spring," he recalls.

In fields without a rye cover, Bennett disks corn stubble in early spring and field cultivates just before planting beans. He also field cultivates soybean stubble once just before planting corn. "If you don't plant right after field cultivating, the weeds come on and it's harder to rely on a herbicide band," he says.

"I'm trying to put the man back into farming, by using time-proven techniques and making them work in modern ways."

Bennett is really sold on fall-planted rye, now that he's figured out the best way to use it with no-till soybeans. A two-year LISA grant helped Bennett and other Rodale Institute farmer-cooperators study the effect of rye on weeds and soybean yields in '89 and '90, but they saw variable results.

Other research using rye in no-till cropping systems has shown that rye suppresses weeds in a number of ways. It supplies mulch, reduces soil-disturbing fieldwork and provides natural compounds that minimize weeds. But rye also can hurt soybean yields.

"In '91, I finally found the right balance in using rye with no-till soybeans," says a satisfied Bennett. Using a rented fertilizer spreader, he broadcasts rye at 2 bushels per acre in late October after disking corn stalks.

"Don't seed the rye too early," he warns. "The earlier you plant rye before fall dormancy, the thicker the matlike growth that results in spring."

Bennett wants enough rye growth in fall for it to at least break through the ground, but he doesn't want heavy or tall growth before dormancy. If you plant early in fall, you'd want to reduce your seeding rate, he says. But a lower rate gives a thinner canopy in spring and early-planted rye would need a higher herbicide rate to kill it.

After planting the rye, Bennett disks again to ensure a good seedbed for spring. He plants soybeans in early May, or as soon as soil conditions and temperature allow. "You want to get in there before the rye is heading — you want the plants tender, so they don't break off or mat down too much. You want as much standing as possible, for shade. So we plant and spray at the same time, to minimize rye knockdown from field equipment."

Bennett's rye is about 3.5 feet fall when he sprays it. "After the boot stage, rye starts taking away soil moisture. But height is a consideration too — you want it to get tall enough to provide shade."

To kill the rye in '91, Bennett used 1.5 pints of Roundup (half the label rate), 1 quart of Lasso, 1.7 pounds of ammonium sulfate and 13 ounces of surfactant, per acre. He started adding Lasso to his herbicide mix in '90, to control nightshade. The ammonium sulfate and surfactant increase herbicide penetration, allowing Bennett to cut his Roundup rate in half and still kill the rye.

"In '89, I killed the rye by disking it in, without any herbicides. But rye alone didn't cut it for me — I had black nightshade come through late in the season," Bennett points out. Disking reduces the weed-control benefits of rye, other researchers have shown.

For superb weed control, Ohio farmer Rich Bennett plants no-till soybeans and sprays fall-planted rye in a single pass.

Bennett drills soybeans at 70 pounds per acre in 30-inch rows, using a Kinze planter. But he leaves the no-till coulters off, and just uses the disk opener. "The coulters would stir up the soil and allow weeds to grow," he says.

Mulch Holds Moisture

"For 30-inch rows without rye, I'd normally no till beans at 50 to 55 pounds per acre. The 70-pound rate covers you in case of poor germination.

"Even though rainfall was low in '91, the rye mulch helped hold the moisture. And 30-inch rows made it easier to cultivate the soybeans. So I decided to eliminate the Lasso in '92."

Almost a month after planting, Bennett cultivates soybeans for the first time, using a Lilliston rolling cultivator. "The rye is brittle then and the Lilliston chops it into little pieces, yet buries very little." The beans are about 8 inches tall at this point, so he can cultivate quickly. Any weeds are about 1.5 inches tall, says Bennett.

"Two weeks later, I go through with a Danish-tine cultivator with the row shovels off, to clear up the row-centers a bit. The beans aren't quite shading the whole area at this point, so it takes an extra cultivation with the 30-inch rows. But that's easier — and better — than working ground in spring.

"The Danish tines also throw rye mulch around the beans. At our field day on August 1, 1991, the soil moisture was just amazing in the rye treatments, while the conventional plots were bone-dry."

Rye provides excellent weed control for Bennett. "I see an occasional broadleaf plus some late velvetleaf, but foxtail, pigweed and lambsquarters are suppressed. The rye also controls volunteer corn.

"Don't use a rotary hoe at all in no-till beans planted in rye. Let the rye do the work, rather than stirring up any weed seeds. That's another reason why I'm moving more of my acres into no till and cover crops — weeds such as foxtail always take off when you stir the soil."

Great Yields, Higher Profits

Bennett had one of his best soybean harvests ever in '91. The rye treatment yielded an outstanding 64.4 bushels per acre, vs. 62.7 bushels per acre without rye. The 1.7-bushel difference was statistically significant. Just getting good bean yields when using rye can be a notable achievement, other farmers have found.

Bennett likes the economics of using rye. It costs about $9 per acre at his seeding rate, but it lets him reduce his herbicide outlay by $39 per acre compared with conventional soybeans that receive a pre- and post-emerge treatment. He will probably continue to use Roundup to kill a rye cover, though. "The only way I've gotten rye to really work the way I want it to is to kill it with herbicide," he notes.

Using Ohio State University custom-operator rates, Bennett figures he saves $15.75 per acre on his fieldwork by using rye. (See chart.) With fall-planted rye, he makes a total of seven passes across the field (including one for soybean harvesting). But that's still two passes *less* than he'd typically use for conventional soybeans, which require extra fieldwork before spring planting and herbicide incorporation. Bennett says his conventional beans usually need two cultivations to control weeds, even with a full load of herbicide. Without a post-emerge, he'd still need at least two cultivations, he says.

Rye also provides benefits that are hard to translate into dollars and cents. "Rye reduces our winter wind erosion, improves soil structure and reduces runoff," says Bennett. It could also be harvested for forage, although he hasn't done that.

In '90, Bennett experimented with planting conventional soybeans into 20 acres of wheat. He was encouraged by the weed suppression he saw. The wheat yielded 50 bushels per acre (vs. 56 bushels in control plots) while the interseeded beans yielded 17.5 bushels per acre. "That's pretty good for a double crop. And I figure we had a 6-bushel wheat loss due to knockdown when planting the beans with an inappropriate planter," he says. Using the corn planter caused too much wheel traffic, he notes.

In late '91, Bennett bought a Great Plains drill to no till beans into wheat. Now, he can harvest even more beans *and* wheat the same year if the weather is favorable. "When I plant rye for harvest, it hasn't been as good a yielder as wheat. I only get 30 to 35 bushels of rye per acre, versus 50 to 60 for wheat," he notes.

He's seen some Japanese beetles on soybeans

Rye Cuts Costs In No-Till Beans

No-Till Beans With Rye		Conventional Beans Without Rye	
Fieldwork ($/A):			
Fall diskings 2x	$15.00	Fall flail-chopping	7.50
Fall broadcasting rye	2.50	Fall chisel plowing	10.00
Bean planting/spraying	14.00	Spring disking	7.50
Cultivations 2x	10.00	Spring harrowing	6.00
		Bean planting/spraying	12.00
		Rotary hoeing	4.25
		Cultivations 2x*	10.00
Fieldwork subtotal:	41.50		57.25
Material Costs ($/A):			
Rye seed	9.00	Bean seed	4.00
Bean seed	5.00	Pre-emerge herbicide	22.88
Herbicide mix	14.00	Post-emerge herbicide*	
Materials subtotal:	28.00		26.88
TOTAL:	69.50		84.13

SAVINGS WITH RYE: $14.63/A

Field steps and material costs based on Rodale Institute network cooperator Rich Bennett's methods and materials. Fieldwork costs based on Ohio State University 1989 Custom Applicator Rates.

*** If using a post-emerge application instead of cultivation, add $18.25 per acre to this column ($24.50 materials + $3.75 fieldwork - $10 cultivations). That would give a total of $102.38 and a savings with rye of $32.88 per acre.**

planted into wheat, but not enough to justify spraying, he says. Bennett hasn't had many pest problems. He avoids corn after corn, so he rarely has trouble with corn borers. "Everybody in our area seemed to have some corn borers in '91, but I avoided spraying. We're fortunate to be far enough north that pests aren't a major problem, especially if you have an adequate rotation," he says.

Bennett hopes to use hairy vetch to control weeds in no-till corn and to eliminate a 10-inch band of Lasso-atrazine. He figures he'll still use a herbicide to kill the vetch cover. But he thinks the vetch will allow him to continue cutting back his use of chemicals in corn, just as rye does with no-till beans.

Aside from supplying nitrogen and soil-improving organic matter, vetch provides next year's corn crop with a good seedbed and enhanced germination. The legume also helps minimize insect pests, as shown by his and other Rodale network trials.

Pastured Pigs Cut Cost

Bennett began testing rotational grazing of gestating sows in '89. Before then, his confinement-raised pigs "never touched dirt from birth until out the door — they were on concrete the whole way," he says. "It took some arm-twisting at first to get me to try rotational grazing, but now I'm convinced there's a lot of potential for increased income there." His results so far indicate that pasturing pigs could cut feed costs and lower labor- and health-related expenses.

To prepare for the first sows in spring '89, Bennett broadcast 35 pounds of red clover seed into a 1-acre grass pasture next to his hog facilities. Using polywire and nylon step-in posts, he divided the pasture into four paddocks. In mid-May, he started rotating six bred sows to a new paddock every two weeks.

"For the first couple of days, I had a hard time keeping the sows in the paddocks. After they learned to respect the fence, it was no problem. I could have doubled the stocking rate, though. The pigs just couldn't keep up with the forage," he says.

Bennett normally feeds 6 pounds of 14-percent protein gestation feed, but cut that back to just 2.5 pounds for the pastured sows. June grass samples tested as high as 17.4 percent protein on a dry matter basis, with July clover samples having more than 25 percent protein. "The sows actually work the grass before they graze the clover. The pasture keeps the sows in good condition too. They don't get too fat," he says.

A hog farmer at heart, Bennett is intrigued by the profit potential of pasturing gestating sows.

After 74 days on pasture, the sows had maintained their average body weight, while those in a confinement-fed control group had increased their average weight by 25 pounds. Both groups farrowed inside, with the pastured sows' pigs having a four-week mortality rate of 8 percent — less than half that of the confinement control group. Bennett weaned 9.4 pigs per litter from the pasture group, vs. 8.6 from the confinement group.

He was encouraged by the first-year results. "Altogether, I figure the pasture was worth about $8 per sow, which wasn't a bad return, considering I had it way understocked," says Bennett.

In '90, Bennett tried grazing 13 gestating gilts, using farrowing "Port-A-Huts" and supplying some grain during lactation. But problems developed. Dry weather slowed pasture regrowth and he was moving gilts to a new paddock every week. Two gilts farrowed early, two died from heat after farrowing in the open and four others chose to farrow under trees rather than use the Quonset-type huts. Pigs that farrowed in the open were moved into confinement.

"Gilts haven't behaved the way I expected them to, especially breeds with a Duroc strain. They have a mind of their own. Their foraging and mothering instincts don't seem as strong as that of some other breeds. Plus some of them broke through the polywire electric fence when they were moved to a new paddock," says Bennett.

Still, Bennett found that pigs farrowed and weaned on pasture using the Port-A-Huts reached marketing weight a full month ahead of those farrowed and weaned in confinement. Lactation feed costs per weaned pig were also lower: $4.44 on pasture vs. $6.03 in confinement.

"There were fewer weaned pigs on pasture (8.4 per litter vs. 9.5 in confinement) but pastured pigs had better weight gains. That's an indication of the healthier living conditions provided by pasture," he says.

'Labor-Free' Pigs

In '91, Bennett selected seven gestating gilts with what he believed to be better foraging and mothering instincts — Yorkshire/Hampshire crosses having less Duroc strain in the breed. He also farrowed five sows in Port-A-Huts and seven gilts in confinement for comparison purposes.

This time, Bennett installed four gates so he could leave the electric polywire fence in place, making it easier to move pigs between paddocks. "That made the pasturing system pretty much labor-free," he says.

The sows proved better than the gilts at farrowing in outdoor huts. Some gilts failed to return to the huts to nurse the pigs, so pig production on pasture from the gilts was lower than for the confinement pigs (4.7 weaned pigs per litter from the gilts on pasture, vs. 7.8 in confinement). The sows on pasture, however, averaged 8.0 weaned pigs per litter.

> *"It took some arm-twisting at first to get me to try rotational grazing, but now I'm convinced there's a lot of potential for increased income there."*

Again, weaned pigs grew faster on pasture than in confinement. From weaning until heading for the grower building, pasture pigs used 1.5 pounds of feed per day per pound of pork gained, vs. 2.0 pounds of feed per day for the confinement group. Feed savings on rotational grazing for the 72-day gestation period amounted to $21.60 per gilt.

Bennett will continue testing rotational grazing for gestating sows. A $1,500 grant from Ohio State University's sustainable agriculture program will help him expand his pasture research. He planned to allow the pigs access to field corn and a protein source in '92. "By letting them eat corn right out of the field and adding some protein to the sow ration while she's on the grass, I think I can increase the birthing and weaning numbers *and* increase the birth weights," he explains.

If he converts more of his cropland to pasture, Bennett figures the return will be worth it. "Considering that cash rents here run $100 to $125 per acre, and you can run 16 or 18 sows per acre of pasture, I'm sure you'd get your money back with pastured pigs. You could even keep the babies there and feed them out on pasture.

"The feed savings is always there on pasture during gestation. You rarely have to provide therapeutic products to pigs on pasture and the manure supplies all the fertility you'd need. But I'll want to see better weaning numbers before scaling up the operation."

Patience Pays Off

Participating in on-farm research has made it easier to enhance the profitability of his farming practices, says Bennett. "Like every farmer, I'm always looking for another effective tool or method to use. You have to be patient, because it can take two or three years to fine-tune your new ideas." Replicated test strips and small field studies allow Bennett to learn more about a technique "before jumping into a new method whole-hog."

If you want to cut your inputs, begin with your fertility program and get an unbiased person to help you, he suggests. "Start with P and K, work your way up to nitrogen, then see where you stand with your soils. As you become more confident, try to do more yourself. Do your own banding, for example, rather than hiring custom applicators. And try to put more life into your soil. Cover crops can help.

"That's the route I've gone. But it probably doesn't do any good to tell people to go slow. They'll jump off the deep end with some things and then slow down.

"In my case, though, I'm now able to avoid chemical company recommendations and many university suggestions too. I'm trying to put the man back into farming, by using time-proven techniques and making them work in more modern ways."

'Balance Is The Key'

Detailed accounting and field research help this farmer fine-tune input levels.

LEXINGTON, Ohio — Despite low prices and poor weather that hurt corn yields from '88 to '91, Tom and Marilyn Culp still made money on their corn acres. Frequent variety trials, good record-keeping and conservative use of purchased fertilizer and herbicides help these grain and hog farmers make the best of their most limited resource: time.

Tom works full time as a traffic engineer for the Ohio Department of Transportation. He farms 330 acres by working evenings and weekends, sharing equipment and labor with a neighboring farmer. "There's really no such thing as a part-time farmer. You work two jobs, one as a farmer," Culp says.

Although he finds tractor headlights indispensable at times, Culp manages to raise corn, soybeans and wheat at or above county yield averages most years. An effective rotation that includes cover crops has allowed him to cut his applied-nitrogen rate by more than 60 percent, virtually eliminate insecticide use and keep erosion below 'T' on rolling silt-loam fields.

Tom and Marilyn Culp
Lexington, Ohio

Started cutting inputs: 1984
115 acres corn
110 acres soybeans
65 acres winter wheat
32 acres alfalfa and hay
5 acres hairy vetch/rye for seed
3 acres sweet corn and pumpkins
330 acres

Livestock: Finishes 400 hogs annually
Growing Season: 157 frost-free days
40 inches of precipitation

Despite maintaining high residue levels and farming 20 percent of his acres as no till, Culp applies below-label rates of herbicides. "With more time, perhaps I wouldn't need any herbicides," he says.

When Culp took over his parents' farm in '78, he and his wife gave themselves five years to see if they could make a go of it. Of the 86 home acres, "not a whole lot was tillable," Tom recalls.

"With Marilyn's brother we were soon renting about 800 acres scattered throughout the county. We were farming a lot of acreage, trying to push the ground for everything it was worth.

Get A Return Or Get Out

"In '81 or '82 we finally realized we couldn't keep putting fertilizers and other chemicals on without getting a return. The maximum-yield scenario wasn't the thing to do. So we started consolidating on acreage. We wanted to ensure that operations were earning their keep.

"Economics was the most visible motivating factor for reducing our use of off-farm inputs. Stewardship was also important," he adds.

To help ensure a return on purchased inputs, Culp soil tests and tissue tests every field at least every other year. "We used to routinely broadcast maintenance applications of 100 pounds of phosphate and 200 pounds of potash per acre," he says. But his fields were testing high in P and K — more than 50 pounds of P per acre and 350 pounds of K. So Culp decided to apply these nutrients only if test values dropped significantly and tissue tests revealed a deficiency.

"We went from being the local fertilizer plant's number-three customer to number 25 in '84," says Culp, who saved more than $20 per acre. Except for some P and K in his starter fertilizer, he hasn't applied any maintenance P or K since '84. His soil tests show K levels are holding stable in the high- to medium-high range. Phosphorus levels are getting high, but aren't excessive. "My bean yields do well even in dry years without added fertilizer," he notes.

Culp has always relied on test plots. "But we were fortunate to associate with the Rodale network in '86 and learn about Dick Thompson's experimental design for on-farm trials. It allows valid research on lowering inputs at a more rapid pace than trial and error, and we're confident in the results," he says. (For the *Farmer's Guide to On-Farm Research,* send $4.95 payable to New Farm Library, 222 Main St., Emmaus PA 18098.)

"We've been doing replicated-trial research projects to help decide the big questions: Should we use a particular product, variety or method? Will something pay its way?"

Thrifty With Nitrogen

Replicated trials showed Culp that his plow-downs of sweetclover and medium red clover following wheat were providing about 90 pounds of N per acre. "Ohio State University only credits 60 pounds of N for a clover plowdown, so we're either doing better or they're being conservative," he says.

"We use a trial-and-error approach to fine-tune some of what we've seen with replicated trials. With corn, for example, we went from applying up to 400 pounds of N per acre in the early '80s to about 170 pounds of actual N per acre since '86."

That's the total nitrogen input, including all the credits from manure, legumes, soil N, starter and sidedress. "We depend on the crop rotation, on cover crops and some manure for nitrogen contributions, in tandem with late-spring nitrate testing. Some of our light-colored soil needs a little more N, so we may boost total N to 190 pounds on those fields," he says.

Culp applies up to 20 pounds of starter N from urea and diammonium phosphate at corn planting. He sidedresses liquid 28-percent in one or two applications, based on the late-spring nitrate test.

Using a five-year average corn yield helps him calculate nitrogen needs. "We've learned to be realistic in our yield goals," he says. Due to poor weather conditions, his five-year average has declined from 124 bushels of corn per acre to about 88 bushels.

Culp thought his crop planning for '92 would show a need to drastically reduce his corn acres, since "four years of tough weather showed us that corn is not a weatherproof crop, and my fixed costs per acre were going up." Thanks to reduced use of off-farm inputs, however, his crop budget for corn showed a positive return even assuming disastrous weather *and* the lower average yield. Instead of shrinking his corn planting to less than 70 acres, he pared it to about 115 acres in '92, down from 141 acres the previous year.

Since he began reducing inputs, Culp's yields haven't changed except for the weather-related reduction in average corn yield, he says. He expected some minor yield losses in beans and wheat, but that hasn't been the case. "That tells me we were putting on more inputs than we needed to years ago. I don't spend another dime on a crop input now unless I'm sure of a return," he says.

Tom Culp bought a 15-foot minimum-till rotary hoe to diversify his weed-control strategy in high-residue fields.

Culp's five-year rotation that includes winter covers helps keep his soybean yields in the upper 40s or low 50s and wheat yields in the 50- to 55-bushel range. "The SCS credits us with seven crops in five years, two of which are meadow crops. Despite 42- to 58-percent highly erodible ground, our soil loss is below 'T' according to our district SCS," he says.

In Year 1, corn follows a spring-plowed clover sod (medium red- and sweetclover). After corn harvest, Culp chisel plows. The following spring, he disks once and solid seeds soybeans. He uses a Danish-tine cultivator/packer ahead of the drill to ensure a 35- to 45-percent residue cover after planting.

"Forty pounds of rye and 20 pounds of vetch per acre are aerially seeded into beans two weeks before leaf yellowing in mid-August. We combine beans and let vetch and rye grow over winter. In the third year, corn is no tilled into herbicide-killed vetch and rye," Culp says.

After corn harvest, Culp chisels in fall. He solid seeds beans the following spring, as before. He uses the Danish-tine cultivator/packer on bean stubble before planting wheat in fall. In spring, he broadcasts into the wheat a 60:40 mix of medium red clover and sweetclover, at 12 pounds per acre. When the wheat comes off, the cycle starts over.

Variety Trials Important

Most years, Culp tests up to 20 different corn hybrids, as many as eight different wheat varieties, plus an equal number of soybean varieties. "I want to see how they'll do in our system before we plant significant amounts," he says.

"We look for indeterminant corn hybrids that will be workhorses — ones that take the nitrogen early and keep plugging away all year; not ones that need 300 pounds of N to perform. In '92, we planted a third of our corn acres with GUTWEIN 2552, a variety that gave us extremely good yields in '91 despite a drought."

He favors hybrids with indeterminant ear length, because "it's easier to strike a balance between populations and moisture, which you can't control. With the determinants, you've got to get the population just right." Culp plants 24,000 to 26,000 seeds per acre the first year he uses a hybrid, then cuts back by 10 percent each year until he finds the best population.

"With beans, we've become better selectors of public varieties. They work real well for us. They seem more adapted to our soil types and our system. If the genetics hold, we'll keep using them," he says.

Because soil types vary in some of his fields, Culp used to test blended varieties containing seed from two distinct lines mixed in the same bag, even for his corn crop. "I figured that for small fields with high, droughty ground as well as low, wet areas, you'd get better results planting blends. But experience showed the blending wasn't worth the pollination variation. Four- to five-day differences — plus drought delays — could mean some late corn didn't get pollinated," he says.

An Acu-Grain electronic combine monitor he acquired as a Rodale Institute network cooperator has made it easier to keep track of yields from small fields and his many experiments. "The monitor gives us yields within 5 percent when we unload, even with partial bins. It's one of the best tools a farmer can invest in," says Culp. (Contact: Acu-Grain, 750 6th St. S.W., Great Falls MT 59404, (800) 648-8007.)

Culp has been cutting his herbicide use since '87, by gradually reducing applications below label rates. "I try to find how low I can go and still have the weed control I need," he says. "I could have used banding to cut herbicide use. But I haven't, due to equipment limitations," he points out.

Good recordkeeping lets him estimate weed-control thresholds where economic crop loss occurs, he says. "I can tolerate some weeds and don't demand clean fields. Our normal weed-control practice now is to broadcast tank-mixed pre-emerge broadleaf and grass herbicides, using a pickup sprayer with 50-foot booms immediately after planting." He reduces rates anywhere from 16 to 50 percent.

"I try to keep herbicide rates low enough that we don't screw up our cover crops. That's especially true for soybeans, as we fly on vetch and rye within a few months of planting beans. Dryness would affect our covers, as would herbicide carryover. We want a good winter cover," he says.

> *"All along, we've been trying to figure out ways to do things better .... Sustainable farming practices have had a very positive impact on our net income."*

This Hoe's For Sale

To see if he could trim weed-control costs even further, Culp decided to compare mechanical cultivation with reduced-rate herbicides. But on some test fields in '89 and '90, Culp found rotary hoeing a disappointing alternative to chemical weed control.

Before buying a rotary hoe, he talked to several rotary hoe users, looked at hoes at farm shows and read all he could find on the subject. Culp figured a high-clearance model would be best, because he prefers to have high levels of surface residue. He was surprised that many of the local dealers and manufacturers' reps couldn't provide enough information about high-clearance models to permit

Need a good used rotary hoe? Call Tom Culp. Wet weather and time constraints have made it hard for him to effectively use a rotary hoe, he concluded.

a realistic comparison. In early '89, he settled on one of his top choices and bought a 15-foot M&W MT 1815 rotary hoe for $3,000.

But he couldn't find any advice on how deep to run the hoe, how fast, how to hoe when the field is too wet but the weeds are at white-shoot stage, how to keep from covering emerged corn plants or how to react to the ever-present rocks. "We had to learn all these practical techniques in the school of hard knocks," he says. As it turns out, the learning curve proved less formidable than the weather.

Culp dedicated 26 acres (corn, soybeans and sweet corn) to a herbicide-free cultivation test in '89. He planned to compare mechanical cultivation to use of herbicides for weed control in other fields.

"Our initial plan was to hoe our corn about four days after planting to kill any weed shoots that developed after pre-plant tillage. To accommodate cultivations, I had increased our planting depth from 1.5 to 2 inches," he says.

"It rained immediately after planting. We were forced to wait five and a half days before hoeing the first time. The soil was still too wet." He ran the hoe at full depth (about 2 inches) at about 8 mph. That seemed to work well, he recalls.

More rain kept Culp from returning to the field until the corn was at the three-leaf stage and weeds had sprouted. "I realized an aggressive hoeing was needed to kill as many weeds as possible, but I didn't know how the corn would take it. I ran the hoe full depth at about 8 mph, which took out a lot of weeds. It didn't seem to damage the corn excessively, although it did cover many leaves with soil.

"The combination of soil-covered plants, continued rain and cold weather killed enough corn plants that I had to replant 80 percent of that field. Another field survived much better because we ran slow — at about 3 mph — for the second hoeing when the corn was at three-leaf stage." He cultivated both fields twice, with the final pass when corn was about 18 inches tall.

Culp liked the way hoeing and cultivation controlled weeds *between the rows*. "But late-season weeds reached significant levels *in the row*." Corn yields were 65.2 bushels per acre for the cultivated fields (including the replanted field of 10 acres) vs. 90.7 bushels per acre for low-rate herbicide fields.

The yield difference wasn't entirely due to weed control, though, because weed counts weren't significantly different between the fields, Culp points out. "Overall yields were far below our previous five-year average of 124 bushels per acre for corn. Extreme weather conditions had dragged yields below that figure for the past three years."

Beans Did Better

A rain respite allowed Culp to hoe soybeans about four days after planting, as he had hoped. "Our next hoeing should have been in about five days, judging by the weed shoots. When we tried to hoe, however, we broke off too many bean plants that had emerged but had not yet developed a true leaf. We had to wait until the beans reached first-leaf stage to keep the damage down."

He did the second hoeing at full depth (2 inches) but at about 3 mph to minimize soybean damage. Then he hoed a third time at the first trifoliate leaf stage at full depth at 8 mph.

He cultivated rowed beans twice at later stages. "The rowed beans were really clean. Weed control was good in solid-seeded beans, which made up 80 percent of the trial. I only saw a slight amount of foxtail. Soybean yields in '89 were 30.9 bushels per acre for cultivated fields vs. 32.2 bushels per acre for low-rate herbicide fields. The difference wasn't significant," he says.

With a year's experience under his belt, Culp decided to continue his cultivation trial in '90, but on fewer corn acres. "I had decided that if wet weather returned, I would use broadcast herbicides where needed."

Corn planting and hoeings went well. "We planted 6 acres of corn and 3 acres of sweet corn early under good conditions. We did three hoeings (at four days after planting at 8 mph; seven days later at two-leaf stage at 3 mph; and four days later at 5 mph) at full depth on dry soil in excellent condition, with good results." Two cultivations followed, as planned, with the last one when corn was 18 inches tall.

"Weed control was very good, with only a few late-season weeds in the plot following alfalfa. In sweet corn that followed mechanically cultivated soybeans, however, I saw significant levels of late-season in-row weeds. Corn yields averaged 73.0 bushels per acre for the cultivated fields and 81.3 bushels per acre for low-rate herbicide fields."

It rained right after planting soybeans, so Culp chose to apply pre-emerge herbicides for weed control. Beans yielded 45.6 bushels per acre.

Cost, Compaction Outweigh Benefits

Cultivation costs compared well with use of reduced-rate herbicides for soybeans, but less so for corn. (See chart.)

"I knew the actual herbicide expenses but had to do some figuring to establish our cost for cultivation. Our actual costs were incurred on too few acres to be meaningful. So I used the Ohio State University 1989 custom-rate figure of $5 per acre for both rotary hoeing and for cultivating of row crops. That cost includes equipment, labor and fuel, amortized over the machine life," he points out.

Culp has a number of concerns about mechanical cultivation. "In our soil type, rotary hoeing and cultivation appear to cause excessive compaction in wheel rows (two rows out of six). We first noticed this compaction in 1989, but attributed most of it to the wet soil conditions that year. In '90, though, compaction was again evident even with dry soil, particularly during the rotary hoeing interval. After we hoed, water stood in the wheel tracks for long periods while none stood on top of the ground in other rows."

Culp also didn't like the yield drop he saw when growing soybeans in rows to accommodate cultivation. "OSU data indicate that 30-inch-row beans average 10 percent lower yields than solid-seeded beans do. Our farm trials agree with that. The yield drop for rowed beans cost us $24 per acre, based on our 42-bushel farm average.

"If you're relying on mechanical means for weed control, more effective rotary hoeing would help control in-row weeds and reduce herbicide use. But I think better *cultivator* techniques would ensure consistency. Because I work off the farm during the day, I don't have much time to cultivate, especially during the hottest part of the day when it would do the most good."

Although Culp hasn't given up on mechanical weed control, he decided to concentrate instead on continuing his reduced-rate herbicide program. "It cuts my costs, causes less soil compaction and fits my schedule better. Overall, it offers a better balance for us," he says. He hopes to sell his rotary hoe.

Thanks to effective rotations, Culp quit using insecticides in his field crops after the '86 season. He went from 45-percent second-year corn in '85 to 9 percent in '86. Eliminating rootworm insecticides lowered his costs by $8 per acre.

Culp does most of his own scouting. But now he relies on it mostly for detection of cutworms, armyworms, aphids, flea beetles, rootworm beetles and Japanese beetles in his high-value crops.

To reduce in-

Putting Cultivation To The Test

CULTIVATION $/A (estimate)[1]	CORN '89/'90	SOYBEANS '89/'90	HERBICIDE $/A (actual)[2]	CORN '89	CORN '90	SOYBEANS '89	SOYBEANS '90
Hoe (3x)	$15	$15	Matl.	$9.85	$12.88	$19.80	$22.19
Cult. (2x)	10	10	Appl.	3.00	3.00	3.00	3.00
Totals	**25**	**25**		**12.85**	**15.88**	**22.80**	**25.19**

1. Estimate $5 per acre per cultivation pass, Ohio State University "Custom Rates Guide," 1989. Culp notes field size and shape, flatness of terrain, labor cost, and width of cultivation tool cause the actual cost to vary from farm to farm.
2. Culp Farms enterprise accounting, 1990. Herbicide cost difference due to 1989 pre-season discounts.

secticide sprays in sweet corn, Culp has been testing pheromone (sex-scent) traps since the '89 growing season. "The pheromones work," he says. "They help you decide when you will need control measures. Based on our experience, we have no need for preventive sprays against European corn borer or corn earworm early in the season."

Culp uses two Scentry Heliothis pheromone traps. (For a retail source, contact Scentry Inc., P.O. Box 426, Buckeye AZ 85326, (800) 548-6531.) He baits one with a corn-borer pheromone lure and the other with an earworm pheromone lure. He places the traps at opposite ends of a 3-acre stand of sweet corn in late May and checks for trapped moths on a schedule that changes with insect development.

"We want to know when actual moth flights begin, to help measure insect pressure," says Culp. Monitoring moth populations with pheromone traps provides him with 7 to 10 days advance notice to decide if a control spray is needed for corn borer, and two days warning for an earworm spray.

The traps are most valuable during the fresh-silking stage, when corn is a tempting target for egg-laying moths. But unlike some pheromone products, these won't control moths through mating disruption. Once female moths are present, there are so many that they'll draw males away from the traps.

Photo Courtesy *Gail Keck,* Ohio Farmer *magazine*

Tom Culp checks pheromone traps regularly in summer to monitor corn-borer and earworm moth populations. The traps help determine if late-season sprays are needed.

Culp saw late-season damage in '89 and '90 due mostly to second-generation corn borers and some earworms. He's rarely had insect problems in early-season sweet corn, but reports that the last two or three plantings often suffer extensive damage.

Each year starting in late April, Culp makes six plantings of sweet corn a week apart. Short- to mid-season varieties are planted first, to take advantage of market windows. Corn borers seem to prefer two of their full-season varieties. "But our observations indicate that in our area, planting dates are more important than variety selection," says Culp.

He bought the Scentry traps for $39.25 each, through a statewide Integrated Pest Management program available from Ohio State University. OSU staff provided a season's supply of pheromone lures for $13. Culp considers the traps worth the modest investment. "They're easy to use, and replacement lures are cheap," he says.

"We've usually seen few or no worms in corn picked through the end of July," says Culp. "And we didn't spray any insecticides in '89 *or* '90, even though corn borers and earworms increased in August both years. But we suffered enough damage, especially from borers, to make 30 to 40 percent of the ears in the last two plantings unsalable.

"We had hoped to be no-spray completely in '91. But we decided it wasn't possible to raise consistent corn throughout the season." When Culp's moth counts for earworm and corn borer jumped on August 4, 1991, he got out the sprayer. He applied Sevin every five to seven days until his sweet-corn harvest concluded two weeks later. "We didn't like spraying, but the traps told us it made economic sense to do so," says Culp, who plans to keep using pheromone traps in his sweet corn.

Culp had hoped to boost his vegetable acreage in the late '80s. He and his wife, Marilyn, knew that effective marketing would be crucial, so they helped organize a high-value crop marketing cooperative of a dozen farmers in Richland County in '88. But it petered out soon after it was formed, when members found they didn't have enough time to invest in marketing.

Time constraints also have slowed Culp's plans to offer farm recordkeeping services to other farmers. "Many farmers lack business skills but know they need to keep track of their margins," he says. Culp feels he could take the system he's using on his personal computer and turn it into a business

"I don't have as many great unanswered questions as I had before," says Culp.

that could help keep neighboring farmers profitable. "The spare time just hasn't been there," he says.

Culp keeps busy in the off season, Marilyn points out. He's very active in his church's youth fellowship program. He also likes to set aside time for special projects, such as spending a week helping Habitat For Humanity build homes for low-income residents.

Hogs Improve Cash Flow

Not wanting their farm income to depend on crops alone, the Culps feed hogs to minimize their financial risk. A cattle-finishing sideline they had in the late '70s and early '80s became a casualty of the Culps' profit-loss analysis.

"We were finishing about 30 head of beef a year. But lacking larger facilities and with no packer nearby, we figured the profit potential just wasn't there. We also didn't like the way the livestock auctions worked," says Tom.

The couple attended Extension-sponsored hog-management workshops. That helped them develop a long-range business plan and a three-year cash-flow strategy.

They bought their first hogs in '81. "We've been making money on hogs ever since," says Tom, who has done a number of feeding-efficiency studies with his feeder pigs. Culp finishes 400 hogs annually. For his hog ration, he used to buy soybean meal and mix it with a vitamin supplement and ground corn. He experimented with growing and grinding triticale and had hoped to use the high-protein grain for up to 35 percent of the ration. But he found he couldn't grow triticale economically.

"Now we use a ground-mix ration that includes soybean meal, vitamins, minerals and some bakery fat by-products. I like cookies and so do the pigs. I just add ground corn to it. It gives good performance."

He experimented with a *lactobacillus* probiotic in '84, but saw little difference in growth or feeding efficiency. "Most research shows benefits for gestating sows, but with just a feeder hog operation, it wasn't helping us," he says.

Culp eliminated the routine use of subtherapeutic antibiotics in '86. "We didn't see any economic benefits in using them. We only use antibiotics the first two weeks after getting new hogs in. Then we cease and only treat sick animals," he says.

Time Will Tell

"All along, we've been trying to figure out ways to do things better," notes Culp, who's pleased with his reduced-input farming methods. "I don't have as many great unanswered questions as I had before, so you could say I'm more into a production mode than an 'R&D' mode with our farming now. And sustainable farming practices have had a very positive impact on our net income."

As with many things in life, you need a balanced approach when reducing inputs, Culp believes. "I'm absolutely convinced that balance is the key. And the balance point is different for everyone," he says.

"All you need is time to figure out what you need to do. Then find the time to do it."

Chapter 2: On-Farm Inputs Pay Their Way

These farmers share their tips for profitable farming with fewer chemicals.

The farmer-cooperators in the previous chapter take pride in being able to show other farmers what they've done to reduce their reliance on agrichemicals. They hold field days, speak at workshops, do on-farm research, participate in sustainable agriculture associations and promote regenerative farming methods in many other ways.

Some of the farmers you're about to meet in this chapter would tell you they're still mastering low-input methods. In a few cases, they told *New Farm* editors in an "Aw, shucks" way that they've simply been pursuing commonsense cost-cutting in their farming methods. That may be, but as these eight case studies reveal, each of these farmers has a lot to be proud of. And all of them are recognized as leaders and role models.

They have much in common aside from the obvious goal of wanting to optimize profits:

- ❖ Willingness to experiment
- ❖ Holistic, "big picture" views of farming for the long term
- ❖ Careful management of on-farm resources
- ❖ An emphasis on improving soil health
- ❖ Knowing the value of good rotations and diversification
- ❖ Making sure on-farm inputs pay their way

Wisconsin dairyman Carl Pulvermacher has been testing seasonal dairying on his award-winning organically certified farm. People are paying attention to grass-based dairying and seasonal breeding, he says.

Virginians Chip and Susan Planck hope to see more "small patch" farmers succeed, as they are, at growing an assortment of high-value crops with next-to-no pesticides. "Chemical-free growing techniques have been around for a long time," Chip notes.

Indiana farmer Vaughn Edwards and Indiana seedsman Kenn Zimmerman say specialty grains and cover crops offer great income opportunities — if you know what you're doing and what your customers need. You can't take too many shortcuts if you want high-quality crops and consistent production, they point out.

A systems approach to healthier soil brings clear-cut rewards, says Illinois grain farmer and equipment innovator Mike Strohm. He describes how cover crops and aeration improve soil structure, allowing you to tap more of your soil's natural fertility.

Grain and livestock farmer Ron Rosmann, a member of Practical Farmers of Iowa, discusses his family's success in slashing chemical costs and controlling erosion. Ridge tillage, composting hog and cattle manure, and on-farm research have helped Ron, who points out some of the drawbacks of intensive hog production.

"Do you really need a silo?" asks Minnesota dairyman Ken Tschumper, who earns a comfortable living from his 100-acre hillside farm and 34-cow dairy herd. Forage production, rotational grazing and low overhead are some of his priorities.

DeKalb County, Ill., farmers Steve, Roger and J.R. Faivre describe their development of information-management systems and improved field equipment. With close to 6,000 acres of corn and soybeans, the Faivres value input efficiency as they strive to minimize problems associated with farming on a large scale.

Plenty of role models here, we think. We know you'll profit from their insights. We'd like to hear about your experiences too. Tell us what you've already done to cut chemicals, and what you expect to do next. Write: What Really Happens, *The New Farm*, 222 Main St., Emmaus PA 18098.

Who Needs Off-Farm Inputs?

Not this organic dairy farmer, whose high ideals have brought higher profits.

LONE ROCK, Wis. — Carl Pulvermacher's second year of relying on rotational grazing reminds him of his second year of pesticide-free crop farming.

"I did well without herbicides in a perfect-condition year in '83, then agonized more in '84 because I felt pressured to do even better. Ninety percent of my forage from April to October '91 came from rotational grazing during a relatively mild year. Can we do it again in '92?" Pulvermacher wonders.

He thinks so. "We'll make it happen. Rotational grazing and seasonal milking require more management, just as it takes added management to reduce or eliminate herbicides. Farming is a hands-on business. With experience, you're successful."

A former management trainee for a nationwide ag-chemical co-op, Pulvermacher used to tell farmers the solution to their pest and yield problems was the convenience of a spray nozzle or an anhydrous knife. No more.

Carl Pulvermacher
Lone Rock, Wis.

Started cutting inputs: 1982

40 acres corn
40 acres barley
20 acres soybeans
50 acres alfalfa, alfalfa-grass
70 acres pasture
220 total acres

Livestock: 45-cow dairy herd

Growing Season: 150 frost-free days
32 inches of precipitation

For Pulvermacher, success in the early '80s came by eliminating virtually all off-farm inputs. He and his wife, Cathy, went from share renting a dairy farm to buying 160 acres and cutting almost all purchased inputs overnight. Dairy manure and alfalfa still provide all the fertilizer they need to grow nutritious feed for their 45-cow dairy.

In '91, home-grown feed, rotational grazing and an organic premium helped Pulvermacher earn a living in dairying, even though conventional milk prices had plummeted by about 25 percent. Today, he still stresses the importance of low overhead and low-cost feed crops for his certified-organic dairy.

"We've got to manage our cost per hundredweight of milk, and invest less time and capital in unnecessary machinery," says Pulvermacher, who likes the adage, 'Think liberal and spend conservative.' He rarely buys new machinery, but figures he's still overcapitalized. "I have less than $50,000 of equipment and it's worth more than I paid for it. I know some dairymen with that much invested in one tractor," he says.

If farmers are going to be competitive in dairying for domestic and foreign markets, they can't continue to buy unnecessary equipment and farm chemicals, he adds.

Farmer To Farmer

Pulvermacher credits three people with convincing him that eliminating chemical purchases made sense. A neighboring dairy farmer, his friend and dairying mentor Dean Swenson, had quit chemicals cold-turkey in the mid-'70s, shortly before Pulvermacher started dairy farming. "Dean had more foresight than the rest of us," he says.

When Pulvermacher began dairying on rented acreage in late '76, however, he was reluctant to follow Swenson's lead in cutting inputs. "I had a supportive landlord, but I was relatively inexperienced," he recalls.

His perspective changed when he and Swenson visited Dick and Sharon Thompson's 300-acre beef and hog farm in Boone, Iowa, in winter '81. The Thompsons, members of the Rodale Institute's Midwest farm network, hadn't used any chemicals since '67. (See "The Last Chapter Hasn't Been Written," later in this book.)

Pulvermacher was impressed with the Thompsons' healthy crops and well-maintained equipment. "That was a turning point. I made up my mind on the way home that I was going to reduce inputs."

Because he spreads 10 to 12 tons of dairy manure per acre on all his fields and plows alfalfa before corn, Pulvermacher doesn't need any purchased fertilizer.

He started slowly, cutting back on fertilizer for a few of his rented acres in '82. He reduced his nitrogen rate by 25 percent on a small cornfield following alfalfa, without any yield difference.

He had planned to cultivate another cornfield without applying herbicides. "After moldboarding, the field turned golf-course green with velvetleaf. I sprayed when corn and weeds were 6 inches tall, before cultivating. I should have gotten most of the velvetleaf with a rotary hoe when they were still seedlings," says a now wiser Pulvermacher.

Determined To Succeed

In November '82, he and his wife bought 160 acres 20 miles away and also rented a 60-acre tract. "The following spring, I really went at it. I eliminated *all* purchased inputs, except for a small amount of herbicide on 8 acres of second-year corn. For those acres, I used atrazine at less than the label rate — just 1 pound per acre, sprayed in a light rain," says Pulvermacher.

Corn yields dropped by about 10 bushels per acre, he figures. "We averaged 90-bushel corn, but the yield reduction wasn't from any lack of fertilizer. I added at least 12 tons of dairy manure per acre, which put *more* manure on that land following alfalfa than I do now. So there wasn't a lack of nutrients such as nitrogen.

"I had reduced or eliminated herbicides without making up for it with better cultivation. At that time, there weren't many people you could talk to about mechanical cultivation without herbicides."

Pulvermacher bought a used John Deere 400 rotary hoe for $1,000 in '84. "The hoe let me keep progressing without inputs. But my first year with the rotary hoe, I remember hoeing 6-inch-tall corn. I was two or three weeks late and didn't know better. Good thing I had a cultivator to bail me out somewhat.

"We had an open house in '85 for the Wisconsin Rural Development Center (WRDC), and you could see foxtail growing within 100 feet of the house. It was embarrassing. The weeds were still out of hand and showed a lack of ability on my part."

After that, Pulvermacher was even more determined to make organic practices work on his farm. "I wasn't going to use pesticides or commercial fertilizer come hell or high water. It was my idealistic values that drove me. With 10 years of experience now, I'd say the dollars are just as important. But I've been chemical-free on the home acres since '84."

He had an alfalfa weevil problem in spring '84, but chose not to spray. "I got the haybine out a week earlier that year. I haven't had an insect problem since then and presume it's because of my rotation."

Pulvermacher typically follows two years of alfalfa with a year of corn, then a spring grain, usually barley or oats. To grow up to 20 acres of soybeans after corn, he usually plows some hay ground that's had only a year of alfalfa.

To maintain soil fertility, he relies on 10 to 12 tons of dairy manure per acre for all of his fields, whether it's hay ground or cropland. He stacks manure in January and February. The rest of the year, he spreads manure daily for sheet composting without incorporating it. On the 33 steepest pasture acres, he manures just the top half with a spreader, and the cattle take care of the rest.

Annual soil tests show that phosphorus levels have risen gradually from the manure applications, while on some of the acres potassium levels have been trending downward. "I suppose I'm going to need potash some day, but my average K value, 233 pounds per acre, has only declined 10 to 15 pounds in 10 years," says Pulvermacher.

Acquiring rotary-hoe experience was the key to successful farming without herbicides, Pulvermacher found. "You have to hoe extremely early. By '86, I was rotary hoeing as little as four days after planting, to see how early I could see a hoeing effect on weeds. If more crop farmers saw the rotary hoe as an aeration tool rather than a cultivation tool, they'd be out there earlier and more on target.

Better Weed Control

"Early hoeing — about six days after corn planting — dries out that first brood of weeds but doesn't eliminate all of them. At second hoeing three to seven days later, you want to catch as many weeds as possible before the two-leaf stage.

"You're really moving with the hoe — I've got to hang onto the tractor. There's dirt flying and I'm concentrating like a stock car driver, covering a lot of ground. I can rotary hoe 4 acres of corn quicker than someone can spray them."

Now, Pulvermacher waits until corn is at least 7 inches high before cultivating the first time, with second cultivation taking place when corn is about knee-high. Timeliness can be a problem, he admits. "I don't do a lot of the cultivating myself. That's usually when I'm making my first hay cutting. I've always farmed with a part-time helper, Ken Schmitz, who has cultivator experience."

Aside from some quackgrass, Pulvermacher doesn't worry much about weeds now that he's mastered rotary hoeing and cultivation. "Quack-grass is a hell of an asset in pasture. You have to manage it in your crops, though. I see quack in the row crops every year, but by second cultivation it's hard to believe I was concerned."

Another turning point came in early '86, when Pulvermacher attended a Rodale Institute "Take Charge" workshop in La Crosse, Wis. "I wanted to find out what kind of cultivator I should buy next, but learned that disk hillers made back in the '40s were still available for my four-row IH cultivator.

"I also remember an Iowa farmer telling me that if you're driving along at 12 miles per hour and see a velvetleaf, you're a year too late with your hoeing. Some quackgrass is OK, but you shouldn't see any broadleaves at the two-leaf stage during your first rotary hoeing.

"Well, I was seeing foxtail take over in corn after corn. Even after three years of alfalfa in another cornfield, there were still some broadleaves. It had little foxtail and yielded about 115 bushels of corn per acre. But I knew I needed to improve my weed control in corn overall, based on field appearances and yields."

Soon after that workshop, Pulvermacher bought a set of used disk hillers for about $80. The disk hillers improved his cultivation by tearing out or covering tough weeds such as quackgrass. "At first cultivation, I have the disk hillers angled slightly to throw a little dirt in the row, but not much. That first time through, I'm using them more as a shield than as a cultivation tool.

"But at second cultivation, corn is growing like crazy. I've angled the disk hillers to throw more dirt on the row, and I'm literally building ridges. Any broadleaves or grasses have usually been beaten back by then, so if you throw a 6- to 8-inch mound of dirt on them you're really stunning them. It looks like ridge-tilled corn when I'm finished," he says.

Economical Yields Are Winners

In the early '80s, some of Pulvermacher's neighbors thought he had flipped out when he chose

Photo by Mike Brusko

"Two of the best kept secrets in agriculture, if you're growing corn, are the rotary hoe and disk hillers," says Pulvermacher.

not to apply any chemicals to his heavily mortgaged fields. "It was a couple years before some of our neighbors' perceptions changed. After two years without synthetic inputs, our yields looked great. People couldn't believe we did it without herbicides."

Pulvermacher's farm also received a lot of publicity in the mid- to late-'80s. With a small amount of funding from the non-profit WRDC in spring '86, Pulvermacher and a few other farmers formed the Southern Wisconsin Farmers Research Network. They began a cost-cutting demonstration. Each farmer planted corn after a legume plowdown, and evaluated corn yields at different nitrogen rates.

Pulvermacher plowed an 8-acre field in which he had grown alfalfa for four years, spread 11 tons of dairy manure per acre, then planted a 98-day hybrid. He applied ammonium nitrate at 0, 50 or 100 pounds of N per acre on different strips, with a separate control plot receiving just 100 pounds of 9-23-30 starter. Corn yields were 124 bushels per acre with zero purchased N, 128 bushels with 50 pounds and 134 bushels with 100 pounds.

"At about 20 cents per pound of applied N, it was a wash. It wouldn't have paid to buy nitrogen," says Pulvermacher. Results were about the same on the other farms, he adds.

In fall '86, Pulvermacher was one of a dozen Wisconsin farmers who, along with the WRDC, persuaded the state legislature to funnel $2 million of oil overcharge funds into sustainable agriculture research. As part of the citizen-advisory council for the Wisconsin Sustainable Agriculture Demonstration Program, Pulvermacher helped channel the money into 80 farm-based research projects from 1988 to 1991. He's also on the 10-member advisory council for the UW Center for Integrated Ag Systems started in late '89.

> *"After two years without synthetic inputs, our yields looked great. People couldn't believe we did it without herbicides."*

In '87, Pulvermacher won his regional PEP$ ("Profits through Efficient Production Systems") contest, a statewide maximum-economic-yield program sponsored by the University of Wisconsin. The only chemical-free farmer in the competition, Pulvermacher grew 162-bushel corn for just $1.16 a bushel. That was the fifth-best of the Wisconsin entries, which averaged 167 bushels at $1.33 per bushel. By contrast, the regional (non-contest) average that year was just 118 bushels at $2.05 per bushel.

Pulvermacher's Extension agent calculated Carl's corn costs at just $1.08 per bushel in '87. It could have been even less, says Pulvermacher. For the PEP$ contest, he was assessed almost 15 cents per bushel for P and K removal costs, based on his corn yield. Contest officials say that was to prevent entrants from applying P and K two years ahead of the competition to avoid getting charged for it.

Since Pulvermacher's manure was free, he was only charged a fuel and labor cost for spreading the manure. But since he hasn't added *any* purchased P or K since buying his farm and his P and K values haven't changed significantly, he figures he shouldn't be charged any P or K removal costs.

"Entering PEP$ was one of the most positive things we did for our farm and for sustainable agriculture in Wisconsin. It really caused a commotion — people had to sit up and take note. People were saying, 'Here's a high-debt farmer who's chosen not to buy high-yield fertilizers and pesticides.' It gave chemical-free farming systems credibility," says Pulvermacher.

He repeated as regional winner in '88, producing 129-bushel corn for $1.27. That was a drought year, when the state's average yield was just 67 bushels per acre. Also by the end of that year, all of Pulvermacher's acres had been certified organic.

In '91, he was back up to 162-bushel corn, at a miserly $1.19 per bushel. He placed seventh of 17 in his PEP$ division. Although now more of the entrants were chemical-free, Carl was the only certified-organic entry.

That year, he also entered the MAX contest, a national economic-yield competition modeled after those held in Wisconsin and Indiana. It's sponsored by the Conservation Technology Information Center and *Successful Farming* magazine. "We did well," says a modest Pulvermacher, who was tops for corn in Wisconsin *and* Minnesota, and in the top 10 percent nationwide.

"Results like that show the power of legumes and dairy manure. There's something good going on here to allow corn yields like ours without *any* synthetic inputs," says Pulvermacher.

He stores ground ear corn as high-moisture corn in a 14- by 70-foot silo, and haylage in a second silo. "My corn is cheap — less than a nickel a pound. And it's good for body condition. With reduced feed costs now that we're grazing more —

and transporting and storing less — I figure we're saving a dollar per cow a day."

Less Milk, More Profit

With his crop program well-honed in the '80s, Pulvermacher decided to improve his pasture use in the '90s. "One of my dairy goals has been to feed a balanced ration at the least cost. But in the mid-'80s, I would have done anything to get another hundred pounds of milk on the rolling herd average. Friends pointed out I was still following Extension dairy recommendations. Reducing inputs hadn't carried over much into my dairying," says Pulvermacher.

"In the late '80s, I was somewhat relieved when I brought the herd average below 20,000 pounds of milk and my name came off the DHIA list. I thought the profitability was still there, because of reduced feed costs.

"I realized we'd been babysitting the cows for nearly 15 years, often putting them in the barn to avoid letting their herd average drop. Now I know we can cut feed costs dramatically for six months of the year through rotational grazing and seasonal milking."

Rotational grazing lets Pulvermacher start his seasonal milkers on pasture as early as mid-April.

Pulvermacher began relying more on rotational grazing of his milkers in spring '91. Before switching, he fed cows haylage and high-moisture ear corn twice a day year-round for as much as 80 percent of their dry-matter intake. "The darn cows figured out they could stand in line for feed rather than grazing in the pasture," he quips.

Now, the milking herd is out foraging from early spring to late fall. In mid-April, they start rotating through alfalfa/bluegrass/orchardgrass pasture divided into 16 paddocks. "There's no way to put up feed of the quality we're grazing," says Pulvermacher.

"Carl's getting exceptional-quality feed off his grazing — and doing it cheaply," says Carl Fredericks, coordinator of the Southern Wisconsin Farmers Research Network. "His forage samples consistently show more than 20 percent crude protein, with 25 to 31 percent early in the season."

The milkers spend up to a day, and occasionally two days, grazing in each 2-acre permanent paddock. Portable polywire lets Pulvermacher move the herd in 5 to 10 minutes. "That's a lot quicker than shaking bedding," he notes.

For his 33 steepest acres, he followed a time-saving suggestion made by Wisconsin dairyman Charlie Opitz and eliminated back-fencing when subdividing a paddock. Pulvermacher just uses a single wire of portable fence. He can start with cows grazing in a half-acre section of one paddock when pasture condition allows. About every 12 hours, he moves the polywire uphill anywhere from a half acre to an acre or more, or eliminates the polywire. "We also laid 4,600 feet of plastic watering pipe and got a portable 40-gallon tank. Cows are always within 250 feet of clean, fresh water now, and I've gotten a quick payback on a $600 watering system," says Pulvermacher.

Dry cows and heifers follow a day behind the milkers in the paddock rotation. "We do a lot less clipping using them as a cleanup crew — usually just one clipping after the fourth grazing. Clipping eliminates thistles and smooths out uneven patches before the next grazing."

While the herd is on pasture, Pulvermacher supplements each cow's grazing with a standard ration of 19 to 21 pounds of high-moisture ear corn and 12 ounces of minerals. Winter rations include 20 pounds of ear corn, up to 6 pounds of roasted beans, and some hay and haylage. Calves get 5 pounds of ear corn and free-choice hay.

Pulvermacher contracts to have his own soybeans roasted, ensuring that all his feed is certified-organic. Most years, he grows at least 10 acres of soybeans and 35 to 45 acres of corn. The minerals are the only off-farm purchase in the feed.

Pulvermacher's hayfields typically average 4.5 tons of dry matter per acre and more than 19 percent crude protein. To establish up to 50 acres of hayfields each year (usually into a barley nurse crop following corn), Pulvermacher seeds a mix of 12 pounds of alfalfa and 2 pounds each of red clover, timothy and canarygrass.

Controlled grazing provides additional benefits beyond cutting feed costs, says Pulvermacher.

Portable polywire lets Pulvermacher move his herd in minutes.

"Rotational grazing's more natural environment pays off in easier breeding and a healthier herd. Old-timers would say that's because cows are off concrete and their feet feel better.

"Confinement dairymen probably think their cows are content. But I'd challenge them to come and see my cows' facial expressions when they're lying in pasture. They're comfortable cows, I'm telling you. Comfort translates into better-producing cows."

Seasonal Dairying Works

Pulvermacher started seasonal milking in '92. To make the transition, he bred 45 cows and heifers in May and June '91 for calving in March '92. "To synchronize our herd, we auctioned 37 cows — some of our best genetics — in early January '92. I figured if we had registered Holsteins we might as well take advantage of that to reduce some debt while improving our lifestyle. We weren't milking at all for seven weeks in January and February, but were back up to milking 45 head by the beginning of May."

Seasonal milking is Pulvermacher's priority now, and he hopes to continue to breed cows on schedule. "Some people have already called to offer to buy any cows that aren't synchronized in early '93," he notes.

He says some dairymen can't fathom seasonal milking, though. "'You don't want to milk cows in winter? Are you lazy?' they ask." For Pulvermacher, it's an opportunity to make the best use of pastures and to use some of the time off for vacation travel with his family. It also gives him additional time to devote to the many sustainable agriculture programs he's active in. It's not unusual for him to take a conference call on the phone in the barn, or to make time for giving a talk on the farm-lecture circuit.

"We need better communicators and more sharing of information," he says. "I'm trying. At the farm seminars I speak at, some farmers can't believe what I'm saying. It takes time to convince some people." Recognizing Pulvermacher's low-input expertise, agronomists at the Michael Fields Agricultural Institute, East Troy, Wis., asked him to accompany them on a July '92 trip to Latvia and Ukraine.

Pulvermacher tells farmers to work their way into new practices gradually. "A hard sale is a permanent sale — it takes longer, but lasts," he says.

Homegrown Profits

With any sort of decent crop rotation on a dairy, the dollars are made by growing your own nitrogen, Pulvermacher likes to point out. The next step is to cut your milk cost by $2 per hundredweight.

Because he started seasonal milking in '92, Pulvermacher expects to bring his milk costs below the $9.99 per hundredweight he had in '91. That figure excludes family labor, debt reduction and depreciation, he points out.

He'll also continue to benefit from the 15- to 20-percent organic premium paid by the marketing cooperative he belongs to, the Coulee Region Organic Produce Pool (CROPP). Pulvermacher is one of nine organic dairymen in CROPP, which has been successful in developing strong production, marketing and quality-assurance programs.

"Making better use of pastures and seasonal milking are about where rotary hoeing was five or six years. People are paying attention now.

"Most of what we're doing was proved 40 years ago. I'm trying to show farmers there's an alternative to the way things are done now. I got into this for environmental reasons. But the reason I get really excited about it is the profitability."

'Be On Time With *Everything*'

Fresh, tasty produce beats bin-busting yields, say these practical, low-input growers.

PURCELLVILLE, Va. — Chip and Susan Planck pay little attention to their vegetable and fruit yields. To them, flavor is more important than quantity.

"We could increase our poundage with more fertilizer, but would rather not," says Susan. "We prefer to have tasty produce to sell."

"We don't really know what our yields are," admits Chip. "But we know if our crops look good, if they taste good and if they're as good as the previous year. We could do yield extrapolations by having check rows and weighing some of the harvest each year. Instead, we know if we're happy with our varieties, our mulch use and our gross per acre."

The Plancks are "small patch" farmers who rely on contouring and intensive growing methods to raise high-quality produce on rolling silt-loam fields. They take a practical approach to everything, from fertility and weed control to crop selection and marketing.

Chip and Susan Planck
Purcellville, Va.

Started cutting inputs: 1980

10 acres sweet corn
5 acres greens and spinach
5 acres tomatoes
5 acres snap beans
4 acres squash and cucumbers
2 acres peppers and eggplant
5 acres other vegetables, herbs and flowers
5 acres small fruits
4 acres melons

45 acres high-value crops; 15 other acres not being farmed

Growing Season: 183 frost-free days
42 inches of precipitation

Growing high-value crops with fewer purchased inputs lets small-scale farmers earn more per acre, the Plancks like to point out. By getting better at managing a small number of acres, they paid off personal loans in the '70s, then repaid a conventional mortgage in just five years in the mid-'80s.

The couple *netted* an average of $867 per acre from 1987 to 1991 from 45 acres of vegetables and fruits at their Wheatland Vegetable Farms in northern Virginia. That figure includes a hot, dry year in '88, when their net income fell to $424 per acre.

"We're making money and are happier not putting on chemicals than we would be putting them on," says Chip.

Experience Helps

The Plancks started farming full time in '73 as junior-partner apprentices with some friends, Tony and Hiu Newcomb of Potomac Vegetable Farms, Vienna, Va. "They believed in growing small-acreage vegetables organically and intensively, and larger-acreage ones such as beans and sweet corn with granular fertilizer and standard herbicides. We learned a lot, and adopted their strategy of growing most crops without pesticides," says Chip.

"For our first season on our own in '79, we used synthetic fertilizer on everything and standard herbicides on sweet corn," says Susan. Despite the couple's acquired skills, their farm's expenses exceeded income by about $4,000 that year.

They weren't discouraged. "We had been told not to worry about losses our first two seasons. In '79, we had great yields in a wet year. It was clear we needed to improve our marketing, not our farming," Chip recalls.

"We did poorly at local stands in our county, and wholesaled scads of melons and handpicked beans. Many of our beautiful tomatoes rotted in the fields," adds Susan.

"We lived cheaply," says Chip. The couple scrounged for equipment at farm sales and lived off market-rate loans from friends and relatives. As an investment in their soil, they also decided to lower their chemical use and improve their fertility program.

Ensuring Nutrient Availability

"We switched our main fertility program to ground limestone and a foliar fertilizer in '80. We also started using horse manure on the intensively grown vegetables," says Chip.

"We use only pulverized ground limestone (calcium carbonate). Based on what we've read and seen, we think calcium is an underrated factor in plant production. While organic matter helps hold water like a sponge, calcium creates more pore spaces. It allows air circulation and water drainage as needed plus better root penetration in a dry year," says Susan.

"Calcium carbonate is more than a soil conditioner. It's a nutrient in its own right and a nutrient releaser that frees up other elements needed for plant growth. So calcium unlocks existing fertility," adds Chip.

"With a pH of 7.3, ground limestone is perfect for vegetables, including potatoes. It's hard to overlime with ground limestone. Hydrated and burned lime, however, have a higher pH and will unbalance the soil."

In '80, the Plancks applied from 4 to 10 tons of limestone per acre, followed by maintenance applications of 2 to 3 tons per acre every other year. They spend about $17 per ton for it now, including application costs. They believe it's well worth the investment for all their high-returning acres. "Our soil structure is noticeably improved and our fertility is up, according to soil tests," says Chip.

The Plancks also use a seawater-based 10-20-10 foliar fertilizer. "It includes food-grade sources of urea, phosphoric acid and potassium, plus micronutrients. It's not organically approved, but it doesn't erode nozzle tips and we have no worries about it being hazardous at all. It's the only manufactured fertilizer we've used since '80," says Chip.

"We dilute it and use a tractor-mounted boom sprayer," says Susan. "We apply 2 gallons of material per acre once every two weeks to *all* of our crops, for up to five applications."

That's equivalent to about 8 pounds of N, 16 pounds of P and 8 pounds of K per acre total, Chip figures. "Since we're spraying a foliar feed that the plants can use efficiently, we're applying much less NPK compared with most dry fertilizers." Foliar fertilizer also makes sense environmentally, because there's no leaching or runoff, adds Susan.

Boom-spraying foliar fertilizer every two weeks is one of their secrets to growing tasty vegetables, says Chip Planck.

The Plancks also use the foliar mix as a starter fertilizer, applied with the planter for all crops except for beans. "It drips out at a steady rate from a cheap homemade attachment. We use it in the transplanter water. We also use it in a backpack sprayer in the greenhouse and on small plants," she says.

"Foliar fertilizer costs us $33 per acre per year. We rely on it as a low-N fertilizer for all vegetables, including sweet corn. We're convinced that the more excess nitrogen you apply, the more watery and less tasty your produce."

Organic Matter Helps Fertility

The Plancks have been spreading partially composted manure and wood shavings on some of their acres since '79. "We wanted to spread as much chicken- and horse manure as we could, but for cost reasons manure hasn't been the critical element in our fertility program," says Chip.

Although small amounts of horse manure are

widely available in their area, hauling it from larger horse barns 18 to 20 miles away costs about $6 per ton. That doesn't include spreading costs.

They've gradually increased their use of manure and have spread at least 20 acres with 10 to 15 tons of horse manure and bedding per acre since '87. At an average analysis of 1-1-1, the manure adds up to 300 pounds of N, P and K per acre to about half the fields each year. That may seem like a lot of fertilizer, especially the nitrogen amount. But less than half of the manure N is available to crops the first year. The bedding's high carbon content ties up a lot of nitrogen. Slow-release N from manure, however, helps the Plancks get good yields whether or not a field is manured in a given year.

The couple also like the way cucurbits, okra and eggplant respond to chicken manure when it's available. But they're careful not to apply it to tomato or pepper plots since that would give larger, bushier plants but fewer fruits, they explain.

Before hay mulch is applied, Susan Planck likes to reduce weeds in some high-value crops using a springtooth cultivator and a Farmall 200.

The Plancks use spoiled hay for mulch on their long-season vegetables, especially the tomatoes and peppers. They mulch about 7 acres per year. "It would be fair, if unorthodox, to call mulch a fertility material, since in addition to helping with moisture conservation, weed control and protection of soil microclimate, it is a significant source of organic matter," says Susan.

"We use at least 10 tons of mulch per acre. We pick up a fifth of it for free, but we usually pay 25 cents a bale, or 50 cents a bale if delivered. We're willing to pay for hay to have a regular supply available, even though that brings our average mulch cost to $350 per acre for 7 acres, excluding spreading costs," says Chip.

Excluding labor, they spend about $142 per acre for limestone, foliar fertilizer, manure and mulch. Since the Plancks only grow high-value crops, they consider that figure reasonable. Although their fertilizer application costs are significant, their purchased-input costs are minuscule compared with university standards for conventional growers, says Susan.

The Plancks could lower their fertilizer costs even further by including more cover crops in their rotations. But they can't always spare the time or the fields. "We like broadcasting red clover into sweet corn with a tractor-mounted spinner just before last cultivation. The clover comes up beautifully," says Susan.

"We always plant 10 to 15 acres of rye in the fall and some buckwheat in the rotation niches, but we're not cover experts. We'd like to do more with covers and mulches, since we've been learning more about the importance of maintaining soil life," says Chip.

Use of appropriate technology can also help control costs, the Plancks point out. They've done a lot of hand-watering from a 275-gallon tank mounted on an old manure spreader, for example. "Applying dollops of water to each transplant during a hot, dry period is a more efficient way of getting water to the desired plant than watering everything with an expensive setup," says Chip.

The Plancks do use some irrigation, however. They rely on overhead sprinklers and portable aluminum pipe, plus some underground mains connecting to hydrants around the farm. They spent about $7,000 in the mid-'80s for their watering system, including the cost of cleaning out some existing ponds and installing pipework. "The system paid us back its first year," says Chip, who might add some inexpensive drip tubing in the future.

Timeliness Limits Weeds

The Plancks spent about $1,000 for chemical weedkillers in '79, when they were farming 60 acres, vs. just $349 for 45 acres in '91. Much of the savings is due to eliminating corn herbicide and discontinuing boom spraying of Roundup to control Johnsongrass.

"We got away from herbicides on sweet corn

after '83. We are disinclined to use chemicals. We assume the less used the better, and have always avoided insecticides by not growing late-season corn," says Chip.

"I did the corn herbicide spraying and I never liked it. You have to empty your sprayer somewhere too, and I didn't want to put dilute atrazine or other sprays anywhere else on our ground. We have many different crops growing and they could be sensitive to chemical carryover.

"After losing money on corn several years in a row when we couldn't irrigate distant fields, we decided to reduce our acreage and stop spraying. We're down to 10 acres of corn now instead of 25, and we just rotary hoe and cultivate it. The weeds are manageable."

Initially, Johnsongrass was a major problem, with infested spots amounting to about 8 acres. The Plancks grew corn or beans there for two years and sprayed Roundup plus other herbicides (Treflan in beans; Sutan for corn) to reduce the Johnsongrass. "That was costly and there are still some Johnsongrass patches. We spot spray or hand wick Roundup now as needed. But we can avoid any problem by mulching on time, so we look for opportunities to put a mulched crop in a Johnsongrass-prone area," says Chip.

The key to low-chemical weed control is timeliness, they believe. "Timeliness is all. Be on time with *everything* — *all* the time," Chip emphasizes. "We've learned that it's well worth having enough help around that you can get into a crop early enough to control weeds."

They use many means to reduce weeds: rotations, transplants, disking, rotary hoeing, cultivating, rotavating, handweeding, hay- or plastic mulch, wicking and spot spraying. They also use a handheld string trimmer in strawberry patches and "for nooks and crannies."

"There's nothing sophisticated about our tillage and cultivation techniques. We usually moldboard plow, disk and cultimulch. Then we use a standard planter or transplanter. We rotary hoe corn and beans once or twice when we think it would help," says Chip.

They cultivate beans and sweet corn using a four-row rear-mount cultivator with Danish tines and an IH Model 200 tractor. For tomatoes and other crops, they often use a two-row cultivator and Farmall 200. The two-row spring-toothed cultivator has some rear teeth that cover the tire tracks.

"We plant on 36-inch rows and keep cultivating

beans, corn, peppers, beets and many other vegetables as long as we can, which is generally three passes at two-week intervals. We cultivate tomatoes once with the tractor, when the plants are 12 to 16 inches tall. Then we walk through with hoes to take care of any in-row weeds and mulch right away," says Chip.

"We use black plastic mulch for 4 acres of melons, as well as a couple acres of early tomatoes and peppers. For these acres, we use a one-row rear-mounted cultivator, a 'cheapo' we bought at a sale. We also close-plant cucurbits, so they take off and blanket the ground," adds Susan.

"We tolerate weeds in crops. We're not perfectionists. We want crops growing and pickable. We'll often handweed once just before picking or as we're picking," says Chip.

"The hardest part about giving up chemicals if you're a high-value grower is doing the weeding right. We've spent more management time on how to forego chemical weed control than on any other area of low-chemical farming."

Fewer Insecticides Too

The Plancks have kept their insecticide use to a minimum through careful rotations and judicious choice of crops. They never make successive plantings in adjacent patches, for example. They're willing to handpick insects once in their chemical-free tomato plantings, after the first generation of potato beetle larvae has emerged.

They use no insecticides or fungicides on tomatoes, peppers, sweet corn, okra, beets, herbs, strawberries, rhubarb, cherries, blueberries and most greens. "We've never had terrible pest problems, and don't think we have to take care of every insect pest we see," says Susan.

In '91, she points out, they spent less than $250 for fungicides and less than $100 for conventional insecticides for a few crops:

❖ "About 75 percent of our bean plantings receive one or two applications of Sevin.

❖ "In a 1-acre patch of squash and zucchini in '91, we only sprayed four rows for cucumber- and flea beetles. We won't spray squash, zucchini and cucumbers past the seedleaf stage.

❖ "We've tried rotenone and sabadilla to control flea beetles on eggplant, but have found only Sevin will work. We spray once a week until eggplant are a foot tall and have blossomed, then we stop spraying.

❖ "To control Japanese beetles on raspberries, we use Sevin on the foliage once each season, before berries form.

❖ "We control aphids in our two greenhouses with Safer soap spray as needed. We usually spray just twice, a week apart, with the backpack sprayer wand held upside down and moved through the foliage.

❖ "We use Sevin on melon plants from germination until blossoms appear, to keep cucumber beetles from spreading bacterial wilt. We use the fungicides Bravo and Bayleton on melon vines from blossoming to maturity, to prevent downy mildew and powdery mildew."

"We've had ups and downs with muskmelons and are willing to use conventional fungicides on them. But we're not clear that they're the difference between success and mediocrity. We produce a lot

(Below) Working in pairs on a Hershey transplanter, the Plancks and their staff set out hundreds of basil plants in less than 30 minutes.

of beautiful melons. We pitch the few poorly netted melons from diseased plants," Chip notes. "With corn, we sell some 'seconds' at market for half price and explain the defects as being safer than a high-spray regimen.

"In a given year for one reason or another, something will do poorly and we can't always figure out why. We expect some setbacks each year. But with successive plantings and diverse plantings, we minimize the damage.

"We prefer to use low-toxicity materials and think Sevin is questionable when you look at oral and dermal LD_{50}s [laboratory measures of acute toxicity based on animal death rates]. We're looking hard at synthetic pyrethroids, but they kill everything and are kind of new, so their safety is still questionable. We're hoping the use of beneficial parasites will replace Sevin as our bean-beetle control. We still have a lot to learn."

Freshness And Taste Help Sales

In '91, the Plancks grossed more than $150,000 selling produce at 17 growers-only markets in Maryland, Virginia and the District of Columbia. "If you're near an urban market, it's easy to sell home-grown vegetables. The marketing comes to us on a silver platter now," says Susan.

Business is brisk every Sunday at the farmers market in Takoma Park, Md., where Chip Planck gets good prices for vine-ripened tomatoes, squash, and other fresh produce grown with few or no chemicals.

In '79, the couple only sold retail at temporary stands in their county and to a few wholesale customers who ran seasonal markets. "We had a marketing plan but not a viable one. Nobody came to our stands. We realized that there wasn't enough business in the county for us to survive, and that we had to develop additional outlets. We learned marketing fast," says Chip. He and Susan are proud that their crops still travel an average of just 50 miles to market.

"We stand or fall on taste and quality," Chip points out. Customer appreciation and loyalty are important, he says. "Many are concerned about how produce is grown. We let our customers know that we are, without question, low-input farmers."

They're always upfront about the materials they use on their crops. "We'll put 'No Pesticides Used' on the sign only for the vegetables where that's true. The bean signs won't say that, nor will ones for melons or eggplant. We don't say we're organic," says Susan.

She notes a number of other marketing factors they consider important:

❖ **Variety selection** — "If it tastes good and has disease resistance, we'll grow it."

❖ **Diversity** — "Selling a wide variety of vegetables over a long period gives us market success."

❖ **Time spent loading and selling** — "It's about a quarter of our labor hours — freshness is primary."

❖ **Having a well-informed sales staff** — "We hire amateurs, usually 16 to 20 college interns who live on the farm from mid-April through October. Not many of them are studying agriculture. But since we all share the hoeing, weeding, tractor work, picking and selling, everyone knows all ends of the operation."

"It's good for the hoer to see the selling end. Our hoers and handweeders become pickers and sellers the same day," adds Chip.

The Plancks always have a crop such as tomatoes that can be retailed *and* wholesaled, or discounted for canning. If a new crop or variety takes too much labor for yields, they stop growing it.

A flock of 100 laying hens also plays a marketing role. "Egg sales are a perfect adjunct for a small farm, and it brings people here with their children to collect their own eggs," says Susan.

Come On Down

Small farms are good for America, the Plancks believe. The couple spend a lot of time teaching farming and encouraging others to go into farming. "We love to talk about farming — we are almost missionary about it and often talk about our operation at conferences," says Chip.

Their standard market handout encourages customers to visit them anytime. The Plancks love it when individuals offer to help on the farm in exchange for farming experience and wisdom. They're not reluctant to share what they've learned, either. Several of their former interns have become farmers.

"A lot of people like rural life, working on the land and seeing where food comes from. Big farms and mechanization hurt that. There aren't enough jobs today that offer outdoor work, especially for women. That's another argument for low-chemical farming, even if it is more labor-dependent," says Susan.

"We have no inclination to buy more land, which is pretty expensive here anyway. We could have added more acres of some crops, but that didn't seem the way to go, either. Instead, we cut back from 60 acres in the mid-'80s by planting less sweet corn and pumpkins and eliminating use of fields too far away to irrigate," says Chip.

"If you want to get into farming, don't worry about huge capital costs — it's not true for vegetable farming. We started with $6,000 of used equipment, total," he notes.

"I think prospective farmers scare themselves. They may want to be organic and are too purist to get going, or they're too nervous about being successful on a particular scale. Being serious about farming is more important than scale or exact methodology.

"Just plant something and get it to market. Be committed to selling your products, and you'll learn marketing in a hurry. Provide adequate fertility, use good rotations and handle minor insect problems when you must."

Be practical and "don't do something unless you have to," he emphasizes, echoing an IPM principle. "Chemical-free growing techniques have been around for a long time.

"Significant environmental damage occurs, both to the soil life and beneficial insect life on the farm, and in runoff from the farm, from overuse of inputs. Judicious use of purchased fertilizers and pesticides, whether 'organic' or 'chemical,' helps you minimize costs and maximize safety for the farmer, consumer and the environment."

Bouncing Back, Farming Better

Specialty grains and rotational grazing brought this farmer fiscal and physical health.

TERRE HAUTE, Ind. — When cropping with chemicals turned Vaughn Edwards into a "cookbook farmer" with heavy debts in the early '80s, he and his wife Donna knew something was wrong. A short hospital stay for heat exhaustion made matters worse, but at least gave Vaughn a chance to reflect on his situation. "I wasn't watching my expenses, and basically had turned into a tractor driver back then," he realized. "I put the fun back into my farming, though. It's an exciting field to be in."

Edwards succeeded by switching to organic methods, by using rotational grazing for his cow-calf herd and by raising specialty grains such as popcorn and spelt. He also rode herd on his equipment budget, while easing into a ridge-till system that helped him cut costs even further.

Vaughn and Donna Edwards
Terre Haute, Ind.

Started cutting inputs: 1984

90 acres soybeans
55 acres popcorn
40 acres blue corn
35 acres wheat
30 acres spelt
25 acres adzuki beans
20 acres hairy vetch
150 acres mixed-legume pasture
15 acres alfalfa
70 acres hay
530 acres

Livestock: 90-cow cow-calf herd of Angus-Hereford X Limousin
Growing Season: 175 frost-free days
40 inches of precipitation

He gives his feeder cattle enterprise plenty of credit for carrying him through the tough times. While studying agriculture at Purdue University in the late '70s, Edwards took over an inbred 100-cow Angus-Hereford X Limousin herd and 250 acres owned by his father, a local doctor. Half the acreage included worn-out fescue pasture. Many of the cattle were in poor condition — "all bone and hide," recalls Edwards.

He quickly culled half the herd, bought new bulls and began culling another 20 percent annually. Soon, he was selling high-quality feeder cattle at a 5-cent premium.

Bigger ... Yet Broker

After graduating in '80, Edwards began farming full time. With old equipment, he limped through two cropping seasons. He grew corn, soybeans and hay on silt-loam soil that had enough clay in it to crust more than he liked after rain.

Then he made a big mistake. In '82, he took Earl Butz's advice to "get big or get out." He rented 180 additional acres, hired full-time help and bought a new, 155-horsepower John Deere 4640 tractor plus a used combine. "We worked around the clock and got 360 acres of crops planted in two weeks," says Edwards. "I was young and foolish, and had good headlights."

He harvested 150-bushel corn and 40-bushel soybeans. But he still lost money. "I was spending $20 to $25 an acre on corn herbicides, and my fertilizer bill was about $55 per acre for corn," says Edwards. Herbicides also gobbled up soybean profits. He used full rates of Treflan, plus Lexone or Sencor and a post-emerge spray. Feeder cattle sales helped trim $15,000 off his debt, though.

He added wheat to the rotation in '83, but a drought that year caused problems for the crops *and* for Edwards: The prospect of harvesting 45-bushel corn and 11-bushel beans for a $45,000 loss took its toll, and he was hospitalized briefly for heat exhaustion. "The air conditioning in the combine cab went out because of a freon leak while I was harvesting wheat on a real hot day. I was dehydrated as well as emotionally and physically drained from all the stress," Edwards recalls. "But I decided to stay in farming, because I knew I wouldn't be happy doing anything else.

"It was a humbling experience. But I didn't know how to farm differently."

He knew what *not* to do: spend more money. By the end of '83, his debts amounted to almost $185,000. He dropped his hired help and 160 rental acres. Then in '84 he cut 50 pounds off his standard 200-pound N rate for corn. He also delayed equipment purchases, buying only a new six-row John Deere rotary hoe in the next three years. That plus an old row-crop cultivator allowed Edwards to get by with spending just $10 per acre on corn herbicides, banding Aatrex and Dual.

"In '84, I met some Christian farmers in Ohio who gave me some chemical-cutting advice. They said I didn't need chemicals for weed control and suggested instead that I rely on rotary hoeing and timely cultivations. Well, with their advice — they were just a phone call away — plus what I heard at Rodale 'Take Charge' workshops, I had real good results," he recalls.

Although Edwards made some dramatic changes in his farming practices, he didn't introduce them overnight, he points out. "I went one field at a time for the most part. You don't want to go cold-turkey when eliminating chemicals. That's stupid.

"We'd drop the chemical fertilizers first and add large amounts of manure or compost to maintain nutrients and organic-matter levels. We'd usually start by manuring a field before planting a wheat crop, then follow wheat with red clover. We would hay the clover cover the same year, then plow it and incorporate more manure in fall or early spring. That ensured maximum availability of nutrients for corn."

Edwards used two kinds of manure: his herd's cattle manure and shipped-in poultry manure. "I had been truck-dumping beef manure in some strip-mined hills nearby, but soon realized that was foolish. I started using it on crop ground close to the barns at 5 to 7 tons per acre. That rate didn't seem to provide enough N release, so I increased it to 10 to 15 tons per acre."

He bought poultry manure by the 22-ton semi-load. "At $16 to $18 per ton and an N-P-K of about 3.5-5-4, poultry manure was comparable to chemical fertilizer in material costs, although it cost a little more to spread. It was dehydrated, pure poultry manure, so I only applied 2 to 4 tons per acre with a lime spreader. I also added some composted manure over the top of some of the permanent pasture and hay ground.

"With proper timing and the right amount of manure, we got the same corn yields as we had with conventional inputs. Manure analysis is important. You want to base your application rates on nutrient value, expecting no more than 50-percent nitrogen availability the first year from manure. You have to keep good field records." Edwards also had his soil tested regularly — every year for his poorest fields, every other year on more-fertile ground.

He cut his herbicide rate in all cornfields by banding 10-inch strips in the 30-inch rows. That reduced herbicide costs by two-thirds, without hurting yields. Corn yields were more dependent on manure management than on the herbicide amount, he observed. As he gained more cultivator experience, he eliminated herbicides entirely in more of his fields.

"Rotary hoe wheels have to penetrate well if you want a good hoeing, I found. So field conditions and weather are important. I would add or take weights off the hoe depending on how the field looked," Edwards says.

"Some parts of Indiana have enough sand in the soil that you

Photo by Craig Cramer

To help phase out chemicals, Edwards banded herbicides on ridges and could spot spray by turning the band on and off from the tractor.

can rotary hoe a day after rain. But our soil had some clay in it, so I had to wait a little longer. I wanted the soil soft and a little wet but not overly wet or rotary hoeing would compact it.

"After two years of manuring, I could coast into the third year of the rotation with fertilizer-free soybeans following fall-planted rye." The rye helped reduce weeds in soybeans. "You don't want to plant too much rye, as it takes too much moisture out of the soil in dry years. Planting a half-bushel of rye per acre worked fine for me. Higher than that was too much."

> *"Even though I was still $17,000 in debt, some pencil-pushing convinced me it would be foolish* not *to convert to ridge till."*

Thanks to his cost-cutting, Edwards broke even on out-of-pocket crop expenses for the first time in '84. But that year he had a pasture problem that sliced into his beef profits. Clover in his best 50-acre pasture had been crowded out by fescue infested with endophytic fungus. Overfertilization and continuous grazing were to blame, he says.

By '85, cattle conception rates were down to 75 percent. And with his cows calving on pasture, Edwards lost almost one-fifth of his calves at birth. "That fescue almost did us in," he recalls.

He made what he felt was the only sensible decision: rotate the field to row crops. It was a tough call, because by fertilizing the pastures he had more than doubled pasture stocking rate, to 1.5 cows per acre. To replace the lost forage, he had to rent pasture 10 miles away.

Converting the infested pasture to cropland proved to be a blessing in disguise, however. It brought in needed cash and increased Edwards' confidence in non-chemical weed control. The field was the last one he planted in '85. The 25 acres each of corn and soybeans received just one rotary hoeing and cultivation — and *no* herbicides.

Even though planted in late June, the soybeans yielded 40 bushels per acre. That's just 3 bushels less than his other beans that year, which were planted a month earlier and treated with Sonalan and a Basagran spot spray. "I cleared $4,000 on those herbicide-free beans, and it was the cleanest beanfield I had ever had," says Edwards.

The fescue sod plowdown in the old pasture helped boost corn yields 10 bushels above his 130-bushel farm average, Edwards figures. All told, he made almost $16,000 on his crops in '85. That plus a cattle sale in January '86 cut his debt to just $17,000.

Ridge Till Makes Sense

Edwards was doing well at cutting costs, but wanted to do even better. With his disk and chisel plow in poor condition, though, he knew his moratorium on new equipment purchases had to end soon.

"I really needed to upgrade my tillage and planting equipment. And even though I was still $17,000 in debt, some pencil-pushing convinced me it would be foolish *not* to convert to ridge till," he recalls. To beat the upfront costs that scare many farmers away from ridge till, he sold his JD 4640 tractor and used the cash to buy a second-hand 115-horsepower JD 4240 and a used six-row Hiniker ridge-till cultivator.

"I knew I'd save a lot on fuel with the smaller tractor and fewer trips across the field. And with the

Photo by Craig Cramer

With rotational grazing, feeder cattle can help lower your risk when reducing chemical use in row crops, Vaughn Edwards found. He and his son Benjamin share some head-to-head wisdom at their family's farm in Terre Haute, Ind.

savings on herbicides with ridge till, I would have been crazy not to switch," he says.

Time was a deciding factor too. With almost 300 acres of row crops to manage by himself, Edwards knew that ridge till would help him get all his crops planted on time. He wouldn't have to plow in fall, so he could keep running cattle on stalk ground and not have to feed as much hay. Pasturing his cattle in spring and summer — and eliminating early summer haying by grazing the hayfields — would free up time to cultivate, which would eliminate the need for post-emerge herbicides. And leasing a ridge planter would save money compared with replacing his conventional planter.

Edwards used the Hiniker cultivator to build ridges in his corn and soybeans in '86. "The wet spring made it hard to cultivate, but I did get in one rotary hoeing and two cultivations on most of the crops. The Hiniker took care of weeds between the rows. But I was disappointed that besides using $8 worth of preplant Sonalan per acre to take care of grass in soybeans, I had to go back and spot spray $2.70 worth of Basagran and Blazer for broadleaves in the row."

Less Herbicide

He was getting serious about phasing out herbicides, though. A new sprayer setup allowed Edwards to turn a herbicide band on and off from the tractor, letting him reduce his rate by spot spraying.

Where he planted soybeans on time in one 25-acre field, he grew 38-bushel beans without herbicides and with just one rotary hoeing and one cultivation. That field was as productive as his sprayed bean fields, and the herbicide-free field was cleaner, he says. This was the same field that had yielded 140-bushel corn without herbicides after plowing fescue sod the previous year.

Edwards did see some ragweed escapes in the unsprayed field. "But a few weed escapes won't hurt you, compared with using full rates and a shotgun herbicide approach."

Late-planted beans in the old pasture suffered an 8-bushel yield loss, but had far fewer escapes. "With late beans, there's absolutely no need for pre-emerge herbicides," says Edwards.

For corn in '86, all he needed was $4 worth of Aatrex plus a single rotary hoeing and cultivation. His fields averaged 160 bushels per acre, from 100 pounds of preplant N provided by red clover and liquid 28-percent, 10 gallons of 11-37-0 starter and 50 pounds of 28-percent sidedressed at cultivation.

His herbicide savings came in handy in early '87, when it was time to get a ridge-till planter. "I leased a new six-row Buffalo that winter, and with the help of some factory reps from Fleischer Manufacturing (makers of Buffalo equipment), my switch to ridging went pretty smoothly," says Edwards.

He banded all his herbicides in '87, with good results. Yields declined a bit overall. Corn yields ranged from 120 to 150 bushels per acre. Beans were down to 30 to 35 bushels per acre, due in part to a dry August. "The weather was perfect for wheat, though, and I harvested 70 to 90 bushels per acre — my best ever."

The '88 drought also hurt yields. "No till and ridge till weren't worth a darn around here. Ridged corn where we were still using chemicals only yielded 70 to 80 bushels per acre, but our corn following a clover plowdown without chemicals made 120 bushels," says Edwards.

He thought the big difference was tillage. With less tillage, the soil structure didn't seem right, especially in low-moisture conditions, he observed. "The soil was simply harder. My ridges dried out quickly because there was no tillage to break up capillaries in the soils. With little rain, the pores allowed moisture to escape. But where we plowed clover, we had a lot of leaf-rolling in corn. It stayed cool inside the whorl — a good indication that even though the corn was moisture-stressed, it wasn't being severely damaged."

His soybean yields ranged from 15 to 30 bushels per acre. "We fared about the same as everybody else. We just didn't have as much invested in the crop.

"All told, though, my crops didn't really respond to ridge till as well as I had hoped until after about four years. I saw some definite yield reductions the first two years, then the yields started coming around in the third and fourth years."

Ridge till also lost Edwards some rental acres, he says. "Some farmers in my area were expanding their acreage and one convinced my landlord that one of my fields didn't look good. Actually, a pipeline company had dumped a hill of dirt in front of the field, and the landlord couldn't see the field from the road. That cost me a cover crop when the landlord took the field back in the fall." Edwards figures he got a good education from the experience, though. It got him thinking about ways to make more with his existing acreage.

Edwards added a year of pasture to his corn-soybeans-wheat rotation starting in '88. "I interseeded wheat with 8 to 10 pounds of red clover (or sometimes 5 pounds each of clover and alfalfa) in February. I took a hay crop — usually 2.5 to 3 tons per acre — in fall after wheat harvest. The following spring, before getting busy with fieldwork, I subdivided clover pastures with steel posts and electric fencing wire. That gave me at least five separate grazing areas along with my permanent pastures."

Better Use Of Pastures

On clover pasture, his stocking rate was one cow-calf pair per acre. He rotated paddocks weekly to allow clover regrowth.

"My cattle also spent two or three weeks on permanent fescue pasture in spring before temperatures rose and the endophyte fungus could start affecting the cows. When we grazed fescue continuously, our weaning weights at nine months were only 350 to 450 pounds. With rotational grazing, weaning weights went up to 500 to 600 pounds, and conception rates jumped from around 70 percent to 95 percent or more. I don't know why I didn't start rotational grazing sooner," he says.

"I used rotational grazing in every pasture, even rented ones. By adding sudangrass in some paddocks or in an adjacent field or two at times, I also avoided a summer slump in forage supply."

Edwards started using portable polywire and a low-impedance controller in '90, making it even easier to add paddocks. "It's easy to Bush Hog a row in pasture and add plastic step-in posts. For the most part, though, I used existing fences and just added a section or two of portable fencing. I usually left the fencing in place all season."

Edwards says he rarely had to buy additional hay for winter feed. "I was normally *selling* hay, thanks to rotational grazing. I did buy some standing alfalfa hay during the '88 drought, though. There's always hay on the market. But good management includes knowing your hay needs for an entire year. I sometimes bought a stand of foxtail from neighbors for cutting hay in late summer if I had spare time or if I had any doubts about having sufficient winter hay."

His cattle adapted well to the ridges when row-crop fields were rotated to forages, he adds. "They just followed the valleys between the ridges, even when they saw me coming into a field." Besides cutting hay costs, the browsing cattle fertilized his fields and kept brush down along field edges. "Plus they were healthier by not being boxed up in mud lots all fall," he says.

He considered dropping either the cattle or the crops, but liked the security of having two complementary enterprises. "As long as we kept them working together, we could be more cost- and labor-efficient. Cattle work peaked during the off-season for grain, and the herd always buffered a poor crop year. The herd really helped my transition to fewer purchased inputs too," he recalls.

Edwards also liked having a year-round cover in his rotation, even though that meant keeping the cattle out of the stubble. In '88, he started seeding rye and vetch after corn in fall. He would aerial seed a mix of one-half bushel of rye and 20 to 30 pounds of hairy vetch per acre. "I used to seed roughly an equal amount of wheat instead of rye, but the rye gave us a little more in the way of natural weed suppression. We also had to sow wheat earlier than I preferred, making it susceptible to Hessian fly damage.

Year-Round Cover

"With vetch, I made sure I treated it properly with fresh inoculant. Otherwise, you wasted your money. I've seen fields where one plant is green and lush, while another plant 4 feet away is yellow and spindly from poor inoculation.

"An old trick I learned from an agronomist when I was a student at Purdue University is to mix the inoculant with a can of soda pop or a molasses-water solution to get it to stick to the seed. I just mixed 2 liters of Pepsi with 250 pounds of seed. With vetch seed at 70 to 80 cents a pounds, it was worth the extra effort. I also took some of the sting out of the price by harvesting my own vetch seed."

Edwards didn't worry about cover crops getting away from him in spring. "We had a rainy spring in '89, and that's exactly what happened with a cover of mostly vetch with a little wheat. But all I did was mow the rank regrowth with a Bush Hog and most of it went down right between the ridges. Where our vetch stand was good, the soil in the ridge was mellow and fluffy. It was just like planting into powder. Where there wasn't any vetch, it was hard and cloddy — more like what you'd expect when planting beans into corn stubble."

He once tried killing a vetch cover crop on some rented ground with 2,4-D about two weeks before planting. "I don't recommend that," he says. "It started raining and the ground was pure mud beneath the vetch residue. If it's wet, leave the vetch in the field to pump out the moisture, and mow it just before you plant. If you don't ridge till, wait as long as you can and disk it two or three times at three-day intervals before planting. If it's dry, it might be a good idea to take out your cover earlier."

Organic Grains Boost Profits

By '88, Edwards was farming 100 of the 240 home acres organically. "I realized in '84 I didn't have the money to spend on chemical fertilizer or herbicides, so that's really when I began 'going organic' with a five-year plan to phase out chemicals. Finances were a big reason.

"Bible study also encouraged my interest in regenerative farming. Biblical research got me thinking about agriculture and church ministry and how there were some discrepancies between church teachings and the written word. I realized chemical farming isn't the best — it's a wilderness for the most part. Conventional farming encourages loss of soil and soil tilth. Chemical companies are wrong when they say they can sustain yields. Use of anhydrous ammonia and soil-compacting equipment isn't the way to go.

Photo by Craig Cramer

Just $4 worth of Aatrex plus a single rotary hoeing and cultivation was all the weed control Edwards needed in his '86 corn, which yielded 160 bushels per acre.

"I didn't want to use chemicals. Nor did my part-time hand like mixing chemicals during planting season. I wanted to do right for the soil. Eventually, I was shooting for the highest-quality organic crops.

"In the late '80s, we kept our same rotation, only we grew popcorn instead of field corn, spelt instead of wheat and clear-hilum soybeans for tofu. We also grew some feed for cattle on those organic acres."

In '89, Edwards quit using chemicals on the other 140 home acres, which were 15 miles away. "We grew regular feed corn, beans and wheat on those acres, planning to switch to specialty grains once those acres were organically certified."

Although seed costs for specialty grains are a little higher, other production costs are about the same as for normal grains, he points out. "The big difference is you can gross 20 to 65 percent more for specialty grains."

He grew VITA spelt, a line developed for human consumption and privately owned by Purity Foods, Okemos, Mich. "It has excellent milling qualities, is 16 percent protein and was grossing $100 more per acre than wheat.

"I started growing popcorn on contract in '87. But at 8 or 9 cents a pound, it didn't pay any more than field corn. So in '89, I started growing gourmet organic popcorn for Indiana Gold, a fledgling company that paid me 16 cents a pound. I got a premium price because the popcorn was organic and I harvested and air-dried it on the ear. It was a little extra work, but I could harvest it with an old corn picker I bought for $500. I dried and stored the popcorn in an old steel corn crib that I converted back from handling shell corn."

Taste The Difference

Edwards liked growing for Indiana Gold because its popcorn was about the only gourmet popcorn grown organically right in Orville Redenbacher's backyard. "I think ours tasted better because it was organic. When you've been on a soil-building program for a few years using manure and cover crops, I think you can taste the difference," he says.

The popcorn added even more flexibility to

Edwards' rotation. When a field came around to the "corn year," he could plant field corn for cattle feed, popcorn to sell or sudangrass for forage.

"Adding pasture to my rotation and having such flexibility with the corn really helped us integrate our livestock and cropping enterprises. To make that integration even better, we improved our manure management to be sure we get those nutrients back into the field. I stockpiled manure on the field edges, then spread 10 to 15 tons per acre starting in May. By then, the manure was halfway composted, the piles had shrunk and the nitrogen was stabilized.

"Our goal was to have a production system that would give consistent results. We wanted it to be less expensive than conventional chemical-based farming, show short- and long-term profitability and be easy for other farmers to adapt into their systems. Some of the things we tried over the years didn't work out. But I didn't let the mistakes discourage me."

"Some things that we tried over the years didn't work out. But I didn't let the mistakes discourage me."

Edwards was increasing his herd size by 10 to 15 percent each year, aiming for 100 cow-calf pairs. He never liked selling weaned feeder calves through conventional channels, though, because he felt he was seldom rewarded for having a better product. "I started selling most of my calves to an organic cattle feeder in '86. I would still finish about five or so a year for sale as halves directly to customers."

He considered starting his own branded-beef product, with the beef to be packed at a federally inspected plant in Indianapolis and sold nearby through a health-food co-op. But for lack of free time, the project never got off the ground.

In '88, Vaughn and Donna established the Indiana chapter of the Organic Crop Improvement Association, a national organic certifying agency. Certification was crucial if you wanted to market organic crops for a premium, the Edwardses believed. "At first, we had a hard time finding the five certifiable farmers needed to start a chapter. Then, people started coming out of the woodwork. We had about 20 members and supporters within a year," says Vaughn, who served as the chapter's first president.

Before long, Edwards was giving a lot of time to OCIA activities as well as helping with OCIA-International programs. He was also involved with the Indiana Sustainable Agriculture Association. "I had to cut back on my time spent with the organizations, because I didn't have enough time for all the farming!" he recalls.

He and Donna also set up a marketing service for certified-organic growers in Indiana. "We were linking farmers with buyers, but only on a small scale. We were providing lists of buyers of certified grain. Running a marketing service can be pretty expensive. I know only a few farmers who've done it. It's not something OCIA does," says Vaughn.

The Edwardses were sponsoring Christian fellowship meetings too. "A bulk of my free time was spent working with people. Ministry had become a second profession for me. I wanted more training," says Vaughn, who enrolled in a two-year program at The Way College of Biblical Research in Rome City, Ind. He expects to be back farming at least 200 acres in mid-'93, when he accepts a ministry position in the Midwest.

"We sold the crops and cattle and are storing the equipment. My father still owns the land. I really enjoy farming and expect to go back into it when I graduate.

"My wife, family and I have proved that biblical principles bring you successful results. You can't test-tube analyze religion, but understanding spiritual laws lets you apply wisdom to secular settings.

"Owning the land isn't necessary," he adds. "We'll succeed in farming wherever we farm. We can't guarantee it'll be organic farming, but it'll certainly be low-chemical."

An Itch To Innovate

Improved soil, new tools and new crops bolster this pioneer's bottom line.

❖

WEST UNION, Ill. — The year Mike Strohm and his wife Darra took over his father's cash-grain farm, a drought withered corn yields to 50 bushels per acre. That disaster in '83 capped a decade of declining yields and increasing input costs on the southeastern Illinois farm.

"It was trial by fire," Strohm recalls. But the experience got him to start managing the rolling silt-loam fields in new ways. Cover crops, an expanded crop rotation and soil aeration helped *double* the farm's return on investment to more than 10 percent in '92, compared with returns of only 2 to 5 percent in his pre-conversion years.

Strohm has improved his bottom line by:

❖ Cutting fertilizer rates by up to 75 percent while holding corn yields at 140 bushels per acre.

❖ Dramatically improving soil structure, which in turn enhances water infiltration, root growth and nutrient uptake by crops.

❖ Increasing soybean yields 20 percent to 55 bushels per acre.

Mike and Darra Strohm
West Union, Ill.

Started cutting inputs: 1984

320 acres corn
250 acres soybeans
64 acres milo
55 acres sunflowers
43 acres vetch (seed)/buckwheat
17 acres set-aside
3 acre pumpkins, vegetables

752 acres

Growing Season: 185 frost-free days
45 inches of precipitation

Diversifying from corn to include soybeans, small grains, hairy vetch, sunflowers, milo and vegetables spreads out cropping expenses and cash flow. The range of crops also increases spring planting and tillage options, while it decreases establishment risks stemming from his region's unpredictable winter weather.

As an Illinois sustainable farming pioneer, Strohm has been called a maverick for pushing new ideas with determination. Yet lately some landowners have offered him more acreage *because* of his unconventional approach.

"Through the '70s and early '80s we were farming corn. We thought we could build up soil fertility levels with fertilizer," Strohm says. "But we never saw any increase. Yields plateaued or went down. Since we've changed directions, we're producing at the farm's true potential while our inputs have gone down considerably."

Recycling Nutrients

"The '83 drought opened my eyes. We lost a tremendous amount of money that year, and I realized we had to do something different. I looked at our farm, and saw there was a lot of time in fall and spring where the land was idle, where we could raise something to benefit the soil and fix N from the air for crops," he says.

He decided to drill two cover crops in fall '84: rye into corn stubble and hairy vetch into 80 acres of soybean stubble.

He no tilled corn into the lush vetch regrowth in May '85. Strohm waited several more days for extra growth, then sprayed a pint of 2,4-D per acre to kill the vetch. He knew he had done something right when he harvested 159 bushels per acre on the vetch ground after only 60 additional pounds of N. That was 23.5 bushels per acre *more* than a field without vetch that received 135 pounds of N. Even counting $10 per acre for vetch seed, he spent $7 per acre *less* on the better-yielding corn following the cover crop. The bottom-line difference in net profit between the two corn plots was $68 per acre. It looked like he'd found a surefire on-farm N source.

"When I started planting hairy vetch, nobody else around here was growing any. I got my seed from Oregon," Strohm recalls. "There was no place to go for information on this kind of farming, but I talked to the older farmers. That's where I got a lot of guidance.

Cover crops of hairy vetch and rye improve soil and boost bean yields for Mike Strohm, West Union, Ill.

"I didn't have to answer to a lot of other people in starting to make changes, but I did have to deal with my dad. He was supportive, but he made me go out and do research — both the ideas and practical application. Now he wouldn't want me farming any other way.

"In '85, I met Terry Holsapple at a Rodale Institute 'Take Charge' workshop in Indianapolis," Strohm says. The two struck up a friendship based on their experimentation with vetch cover crops in similar growing conditions and a desire to replace off-farm inputs with on-farm resources. They still trade experiences and speaking engagements on each other's farms for field-day programs.

In fall '85 Strohm aerially seeded 25 pounds of vetch per acre on soybeans just before leaf drop on one field and drilled 20 pounds per acre into bean stubble after harvest on another. Dry weather caused poor germination of the flown-on vetch, and he didn't repeat the attempt. Drilling vetch gives much better results because of the assured seed-soil contact, he says.

Besides the N savings that he needed to push his cropping system toward profitability, Strohm found his fields were holding soil much better with vetch. His county Soil Conservation Service rated his '85 soil loss in corn following soybeans and vetch at just 650 *pounds* per acre. The local average for corn following soybeans is 5 *tons* per acre.

Even so, as he looked more closely at his soils and fertility, he learned from a local banker that he was still losing $10 in nutrients per ton of lost topsoil. Those numbers help him put dollar figures on his soil saving and further justify his rotation and cover-cropping practices. By cutting soil loss 5 tons per acre, he saved about $50 per acre in lost nutrients. Previously, soil loss had ranged up to 7 tons per acre. His T level, the estimated rate at which soil loss is offset by natural formation of new topsoil, is 4 tons per acre.

Vetch Switch Works

Within several years, Strohm observed that southern Illinois winters weren't consistently friendly to fall-planted vetch. If the stand did survive ice storms and repeated freeze/thaw cycles, rank regrowth often caused a problem. In late spring the fast-growing legume often sucked out too much moisture before he could kill it without a planting delay — which proved difficult. Mowing didn't work for him. His only vetch-controlling options seem to be 2,4-D or tillage, he says. Fitting in fall seedings around harvest also proved a struggle because of scheduling, labor and machinery demands.

He took a step back to analyze his situation. Strohm redesigned his cover-crop management to capture vetch's benefits without its problems. He decided to plant vetch, which is considered a winter annual, out of season and kill it in the fall. By adding winter wheat to his rotation, he opened up an early-summer planting window for vetch. He reasoned that after four months' growth, vetch till-killed in fall could supply enough residue to be a cover *and* make an N contribution the following year.

He likes his summer-planted vetch system, refined over the years to feature:

❖ Harvesting small grains in late June.

❖ Controlling weeds and grasses in the grain stubble by disking or using a burn-down herbicide. To maintain surface residue, Strohm uses Ingersoll conservation disk blades that are notched and slotted.

❖ Sowing vetch in 8-inch rows with a Tye drill at 25 pounds of seed per acre. The seed he sowed in summer '92 was the eighth generation from his original seed. The Strohm-line vetch has better germination, grows more vigorously and is generally better adapted to his growing conditions than the out-of-state seed.

❖ Watching the vetch put on significant growth with late-fall moisture.

❖ Disking the vetch about November 1 or before major fall rains as one of the season's last field operations. The tillage kills about half the vetch and opens up the soil enough to improve surface-water infiltration. Winterkill leaves little vetch to regrow in the spring.

❖ Checking nodulation in late fall to evaluate in a general way the potential N contribution. He hopes for nodules the size of quarters.

❖ Using a Hach soil-test kit several weeks after planting corn in spring to check N contribution from vetch and to determine sidedress rates.

To track the success of his inoculant, Strohm grubs roots of developing soybeans to check nodules for the juicy, reddish internal appearance that signifies active nitrogen fixation.

Water is critical to his summer-planting success. To ensure sufficient soil moisture for germination, Strohm waits after grain harvest to sow his cover-crop vetch until at least an inch of rain has fallen. Growth is slow through the summer. "Vetch doesn't like the heat, but it makes it through," he says. With the several inches of rain per month his farm usually receives, the vetch develops roots even though little top growth takes place until its fall spurt. "Vetch will just sit there, then explode with a foot of growth in two weeks' time. It's usually more than knee-high by disking."

Strohm is after green matter in this application, not necessarily a pure stand of vetch. "A few weeds or a little volunteer wheat won't bother a thing. It's all good organic matter. We just roll it back into the soil come fall."

"This is nature's most perfect nitrogen storage," he says. "There's little nutrient loss, because there's almost no decomposition below about 60 degrees. Vetch-residue nutrients are dormant until spring heat reactivates soil life. Synthetic N won't hold without leaching."

Vetch-N release is easy to track. Before planting, Strohm's soybean stubble fields test at 3 to 5 parts per million N, while ones with a 50-percent residue cover of killed vetch measure about 5 to 10 ppm. Within a month, warm temperatures jump-start soil life. That reduces vetch cover to about 25 percent and mineralizes N to the 25- to 30-ppm level.

"Vetch is a scavenger for potassium, too. In spring, the potassium from the plowdown is 100-percent available, while 50 to 60 percent of the N will be available. Two tons of vetch biomass, at 4 percent N, puts about 160 pounds of N and about that much K into the soil. This is important for me, because I've got a high level of potassium that isn't available otherwise," Strohm explains.

The summer-vetch system supplies all but about 60 pounds of the N he needs to produce 140-bushel corn. At planting, he applies about 30 pounds of N per acre from 28-percent liquid with 1 pint of 2,4-D and a small amount of residual herbicide. After the Hach test tells him how much N is available, he adds the balance needed when the corn is 4 to 6 inches tall. That's when he makes a pass with a subsoiler of his own design that is a major part of his soil-health system.

Farming Deeper

"Fertility represents only 5 percent of your crop," Strohm says. "The rest is air and water. Tillage keeps you working only the top 4 to 6 inches of soil. How much air and water can you store in that?

"Our soils in southeastern Illinois are unglaciated and have a heavy clay underlayer. Organic matter is low, often at about 1.5 percent," he explains. "I had problems with compaction, which I saw as a symptom of bigger problems: soil that was so tight it prevented infiltration and good root growth."

While the vetch worked well on top, Strohm wanted a way to expand the active soil layer that would support plant growth. He wanted an efficient tool that worked as part of a complete soil-health system. It needed to improve overall soil condition without diminishing the protection afforded by surface residue.

In his farm shop, he fabricated a new tool he dubbed the Strohm Row-Crop Aerator. The tool features straight, flat, forward-angled shanks on a heavy-duty toolbar. Five-inch wings mounted at the bottom of the shanks fracture subsoil along natural fault lines.

Strohm first used the aerator in '84. During that drought year, aerated fields gave him 126-bushel corn yields, a 33-bushel increase over non-aerated fields. He knew he had something that would make a difference in crop production without yearly chemical inputs.

"The aerator has helped my crop roots reach all the soil they need," he adds. "It's also evened up the drainage in my fields. The low places don't show anymore.

"With a normal subsoiler, you have to go back and rip up the field again the next year. From my own experience, I recommend aeration 12 to 14 inches deep in the spring, when corn is 6 to 10 inches tall," Strohm says. "It's important enough that I'll stop planting beans to run my aerator."

The aerator leaves a loosened vertical channel down to subsoil with little disturbance on the surface. "I'm not mixing soil — just cutting it, breaking it, then letting the corn plants stuff the slot full of roots," Strohm explains. Digging with a shovel in aerated row middles shows extra root growth toward the passageways. When he sees those white shoots tunneling through the soil, he knows he'll be seeing yellow — and green — later. "A pound of roots gives a pound of corn," he says, quoting agronomic consultant Dr. Bill Becker, Springfield, Ill.

To increase benefits from the subsoiling pass, Strohm outfitted the aerator with hoses to apply additional N. He uses 28-percent or a pH-neutral material that is conducive to root development. Using 60-inch shank spacing on the toolbar allows every-other-middle placement and quicker field coverage.

Opening up the soil profile in spring, rather than in fall, has its benefits. Soil is warm enough to be biologically active. Adding just the right amount of fertilizer, at root level, has a synergistic effect, according to Strohm. "It's a case where one plus one equals three."

The Strohm Aerator is a subsoiler developed on the Strohm farm to deepen the biologically active soil layer.

He uses the subsoiler only during the corn year in his rotation, which may be every two or three years. Benefits from the passes, however, are cumulative. The most dramatic evidence may be how fast his fields soak up rainfall. In '91, his soils had average infiltration rates of 3 to 5 inches of precipitation per hour, while the average for neighboring farms is only one-half inch.

That figure is much more than an enviable milestone in soil improvement. "In '90, we had a heavy rain just before a dry summer. The water our field absorbed translated directly into extra yield, compared with the fields around us," he says.

His farm staff provides production and field testing for improved versions of the aerator, now used throughout the Midwest. (Contact: Strohm Brothers Inc., R.R. 1, West Union IL 62477, (217)

279-3306. Twenty models available in '92 included a $3,500 six-row unit and a $10,000 12-row unit.) The implement business involves two of his brothers who contribute their expertise in law and marketing. He's comfortable with the farm-based enterprise, despite the demands of managing a growing corps of employees. It fits with his own efforts to use home-grown fertility more efficiently so he can purchase less.

"If my soil goes anaerobic, which it does in a straight corn-bean rotation on strict no till, my pick-up efficiency from commercial fertilizer drops sharply from the normal 50-percent rate. By creating a 12-inch aerobic zone, I'm increasing the size of my farm by deepening the nutrient well," he says.

Strohm sees the aerator as a tool that will help him — and other farmers — develop a long-term regenerative farming system. He's introduced many of his customers to a systems approach that centers on dynamic soil life. Some become interested first in aeration and some in planting hairy vetch. Eventually, they hear his whole soil-based plan to regenerate Midwestern soils above- and below ground.

He's a client of agronomist Becker, who operates Central Illinois Agricultural Research Farms, an agronomic consulting service. "Mike's soil is showing improvement in about all facets, even in organic matter where we didn't think we could do it," says Becker. "He's cut his soil loss to zero and now is building soil, thanks to aeration and cover crops. He's got more earthworms, greater aerobic life and better root systems."

Lab results document the following soil changes between '87 and '91. (During this time Strohm used cover crops and sidedressed N, but no maintenance fertilizer.)

❖ **Organic Matter.** Increased by 36 percent, from 1.6 to 2.2 percent of soil by weight.

❖ **Cation Exchange Capacity.** Increased by 7.5 percent, to a level of 12.9 percent. CEC is a measure of how well soil and roots can exchange positively charged nutrients (cations) such as potassium, calcium and magnesium.

❖ **Phosphorus level.** Up by 15.5 percent in the top foot to 67 pounds per acre.

❖ **Potassium level.** Holding steady at about 315 pounds per acre in the top foot.

❖ **pH level.** Increased slightly, from 6.4 to 6.7.

Hard numbers earn Strohm a hearing as an advocate for sustainable farming. "When you can influence production 10 to 20 percent, you have a direct change on a farmer's wallet. Instead of going for more yield, I work for more efficiency," he says.

Finessing No Till

Cutting back on pre-plant tillage is one way Strohm increases his margin, though it means his herbicide use has remained fairly constant in recent years.

The wet fall in '90 was an exception. It led him to do a good bit of disking in the spring of '91 to prepare for corn and beans. The tillage took care of early weeds, and enabled him to direct dollars to labor rather than to chemicals — a distinction he appreciates. "When that money is spent for herbicides, it never comes back here to Clark County. I can pay a guy for one disking, and I have money circulating in *my* community. Plus, I still get some erosion control and my infiltration rate is maintained."

With pre-plant tillage, Strohm only needed a single post-emerge broadcast application of Pursuit in his soybeans. By using a 75-percent rate it cost only $15 per acre.

In '92, Strohm no tilled most of his acres because he needed greater erosion control. "Some places you *have* to pure no till, and some people will tell you it's the only way to go. But it's just one production method. You have to do whatever it takes to make things work. That's the key to survivability," he says.

Weed-suppressing rye cover crops have also helped Strohm contain herbicide costs. For several years, he was using BALBO rye as a cover crop before no-till soybeans to provide weed control for as little as $13 per acre. He learned in 1991 he could disk down shoulder-high rye, hit it with his Aer-Way residue-management tool, then plant soybeans with his Tye no-till drill. The Aer-Way has flat, ground-driven 8-inch spiked teeth on a heavy mounted shaft. Adjustments change the pitch and depth at which teeth penetrate the soil, while leaving residue on the surface. (The Aer-Way is available through Strohm Brothers or from Holland Equipment Ltd., P.O. Box 339, 20 Phoebe St., Norwich ON N0J 1P0, (519) 863-3414.)

To control weeds in his '92 no-till corn, Strohm used a full-rate application of Princep and atrazine in two field passes: half the material sprayed 60 days prior to planting and the balance at planting.

He used a versatile new multi-crop planter of his own design to seed soybeans on 8-inch row spacings. Cool weather slowed development, but the planter gave him a perfect stand in Roundup-killed rye residue, while the spacing gave him an early canopy for weed suppression. A two-thirds rate of

Pursuit went on after planting.

He made two disk passes in fields where he had green-manure hairy vetch in spring '92. Besides saving on herbicides, incorporating vetch residue into the soil retains 50 percent more N than leaving it lie after a chemical burndown, Strohm says. Less N volatilizes and more is digested by soil organisms.

Finding more cash crops that fit into his sustainable farming future is a priority for Strohm, as well as a personal challenge. "I just love to try something new," he admits. In recent years he has grown as many as six different grain crops in one season. In his rotation, soybeans follow corn. In the third year he plants sunflowers, milo, seed vetch or wheat. Rye is his cover after corn, and hairy vetch fills in after soybeans and wheat. Buckwheat can stand in when overwintered vetch intended for seed harvest freezes out. He's flexible in responding to soil condition and weather, but always keeps a growing cover or residue over winter.

Strohm needs cover for erosion control, even with no till. But his new cash crops contribute more than just dollars to the farm. "Every time you put a different plant in your rotation, it has a mellowing effect. Better soil life comes along with it," he observes. "It's hard to pinpoint, but buckwheat leaves the soil smelling different than vetch does. Too many people don't take the time to get off the tractor to smell their soils and find out what's going on."

Golden Harvest

Strohm looked closely at a trial crop of sunflowers in '91, and he liked what he saw. He planted 24 acres of Cargill 102 sunflowers May 28 in 30-inch rows with 100 pounds of 28-percent N (50 at planting and 50 at aeration). He planted 26,000 seeds per acre, resulting in a final population of about 23,000. Broadcasting 1.5 pints of Treflan the day before planting, followed by a single cultivation, left "not a weed in the field."

Strohm says his normal row-crop combine header worked well for harvest. The sunflowers were ready just before soybeans, helping him spread out harvest chores.

He waited three weeks for stalks to dry down, then hit them with the Aer-Way. Before planting the field to corn in spring '92, he used the Aer-Way again and harrowed.

His analysis of the new crop illustrates how broadly he thinks now in terms of farm systems. "When you're farming, you've got to take *everything* into consideration. Lots of people don't do that nowadays. They just say, 'I'll plant corn and soybeans and see what looks good next year.' But you've got to plan three years down the road all the time to make this system work.

"Sunflowers left the soil really mellow, seemed very allelopathic to all weeds, performed well in a dry year, needed no special equipment and required no change in farm practices. They look like they will fit very well in a rotation, with a net profit about like 55-bushel soybeans," he says. "The main thing is finding a dependable market."

Strohm sold the crop for 12 cents per pound to a bird-food processor. He harvested about 2,500 pounds per acre, 1,000 pounds per acre above his breakeven figure. Because the USDA classifies sunflowers as a minor oilseed crop, Strohm could sell the crop and still receive deficiency payments for placing the field in the 0/92 program as part of his wheat base.

Beyond that current economic incentive, Strohm sees additional potential in sunflowers. "If you get a good stand, you could raise sunflowers organically. And with a good cover crop of vetch, you could produce them with no fertility inputs."

For '92, he more than doubled his sunflower acreage. He planned to evaluate the crop's response to different sources of naturally fixed N, comparing the sunflowers' performance in fields following soybeans with fields where the previous crop had been vetch for seed.

The 'flower seeds again went to birdfeed. He moved up harvest of the 90-day crop to near Labor Day thanks to an early May planting.

High-Value Cooperation

Strohm's itch to innovate doesn't stop with machinery or cropping systems. He and high-value grower Roy Cork started raising organic vegetables in '91. Cork raises, harvests and markets vegetable crops for several area farmers in exchange for use of their land, soil preparation and planting. After a successful year in '91 on 2 acres, his partnership with Strohm expanded to 3 acres of sweet corn, tomatoes and Atlantic giant pumpkins in '92.

Cork sells the produce from his front yard in the nearby village of West Union, population 300. Strohm says good traffic on State Route 1 and visitors drawn by two state recreation areas nearby provide strong demand that makes direct local sale possible.

"We don't press the word 'organic.' People buy our vegetables because they taste better. I can really tell a difference," he says. "And it's a partnership between friends that's developed good fellowship and kept money in this community. Whatever money I can keep here changes hands five to seven times. I like that."

To plug into the regional farm economy, Strohm bought used seed-cleaning equipment to begin selling Illinois-adapted vetch seed to area farmers. He says Illinois is a vetch-deficit state that produces 50 to 70 tons of seed a year and still has to import stock from Nebraska and Oregon. He harvested 10 acres for seed in '84 and was up to 40 acres in '91. But winter freezes wiped out 85 of the 150 acres intended for seed harvest in '92 and left a 75-percent stand on the balance.

Strohm nets about $300 per acre from vetch seed on a harvest of 500 pounds per acre. With no promotion other than a reasonable price and his personal advocacy of vetch benefits for erodible soils, demand consistently exceeds his production.

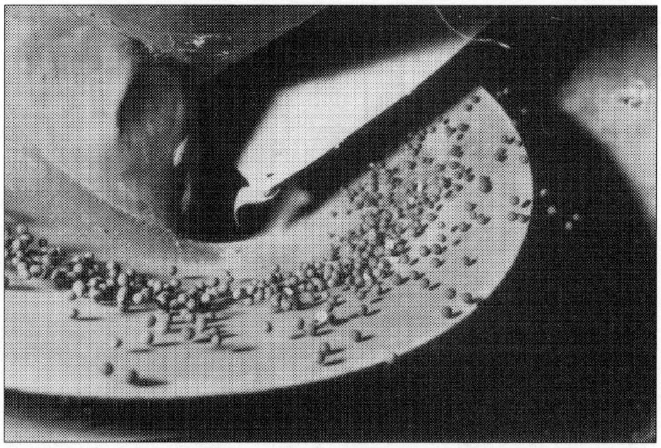

Selling vetch seed brings income for Strohm and regionally adapted strains to his farmer-customers. A spiral-type seed cleaner is part of his on-farm processing system.

Vetch planted for seed harvest goes in with rye at a rate of 20 pounds of vetch per acre to 60 pounds rye. Using Poast or Fusilade herbicide in late spring ends rye competition while preserving its trellising benefit. Spraying also gives Strohm a cleaner run of vetch seed, he says, by keeping weeds and rye from heading out.

In presentations throughout the Midwest explaining his experiences with soil-building and farm-enhancing techniques, Strohm has given innovative farming a high profile. The Illinois Association of Soil and Water Districts named him the state's 1992 "Sustainable Farmer of the Year." The honor acknowledged his improvements to soil and net profit on his own farm and his advocacy of sustainable farming concepts throughout the state and region.

Strohm doesn't allow himself to think he's got enough of the right answers just because a new idea works for a year or two. If a weather shift doesn't change the economics, he's always looking for a better combination to further improve soil, profit, sustainability or his contribution to Clark County's economy. With a system like Strohm's, these benefits go hand in hand. Now that's something he can smile about.

It Gets Better Every Year

Rotations and tillage topple weeds, while savvy marketing triples this organic seedsman's net.

CRAIGVILLE, Ind. — When seed distributor Kenn Zimmerman started selling cover-crop seeds to organic farmers in '85, he thought his customers were crazy. "I saw them as a ready market for vetch, rye and clover seed. But I just didn't think there was any way you could grow organic grains profitably on a commercial scale," he recalls.

Now Zimmerman admits he jumped to the wrong conclusion. "While selling those cover crops, I saw enough 600- to 1,000-acre organic farms to convince me that I could do it, too," he says. By '92, all the fields on his family's 154-acre farm were certified-organic, or in the three-year transition required for certification.

"It didn't take a Rhodes scholar to figure out consumers aren't terribly fond of heavy chemical use. And there are a lot more of them out there voting than there are farmers," he says.

Kenn Zimmerman
Craigville, Ind.

Started cutting inputs: 1987
- 50 acres open-pollinated and specialty corn
- 50 acres clear-hilum soybeans and adzuki beans
- 50 acres spelt or other small grains

150 acres

Growing Season: 160 frost-free days
36 inches of precipitation

After deciding to switch to organic methods, Zimmerman proceeded cautiously, converting a field or two each year. His rotation closely resembles the corn-soybean-wheat sequence common in the area. But instead of low-paying commodities, he grows open-pollinated and specialty corn, clear-hilum soybeans, adzuki beans, spelt (a winter grain used in health-food pastas), hairy vetch and other unique crops. By selling his harvest as seed stock to farmers and food processors through his company, Fruited Plain Seeds Inc., or tapping other high-paying markets, he doubles or triples his per-acre net compared with conventional crops.

"That essentially doubles or triples the size of my farm," he says. "We can't afford to buy more land. So we have to find ways to make more from what we've got."

Finding The Niche

While on the road selling cover-crop seed, Zimmerman noticed that many of his customers grew specialty grains and beans for the organic market. "So I asked the growers where they bought their seed, and they'd say so-and-so has been saving it for decades," he recalls.

A light went on in Zimmerman's mind. "From my years in the seed-selling business, I knew that when farmers save seed like that, they usually end up with low-quality genetic stew unless they're extremely careful," he says. "I realized there was no source of supply for high-quality seed for these crops. And I knew right then that I had found my niche."

The niche of selling seed directly to farmers has been slow to develop. "We'd starve if farmers were our only market," quips Zimmerman. But food processors have been quick to take advantage of his seed-producing skills. They buy the majority of Zimmerman's harvest for their growers to plant, recognizing that his genetically uniform seed stock is the first step in turning out high-quality final product.

Zimmerman provides the extra hours and hard work required to meet the high standards expected of a seedsman. Except for roguing off-types, there's little extra work growing seed crops, he says. "But once you get to harvest, it's a different story. First, you have to do a thorough job cleaning *everything* — combines, wagons, elevators, etc. — to keep from mixing crops." Zimmerman uses his air-screen and spiral seed cleaners to remove weed seed and any other contamination. Then he grades and sizes seed before bagging his final product.

"People complain about the high cost of seed,"

he observes. "But they'll find out what a bargain it is if they crawl around in a combine for three or four hours just to get it clean. There's no easy way to do it."

Even if you have the patience and skills to grow and clean organic grains or seed crops, it's also essential to have a flair for selling, says Zimmerman. "Most people come to organic farming from the production side. But marketing is everything," he observes. "My primary expertise is marketing. In the seed business, first I learned to sell, then to buy and then to grow. It's the same for organic grains. First, you need to find out what you can sell, then you find a way to produce it. If you can't produce it, you find someone who can, and maybe they can do it cheaper than you can do it yourself."

Indiana seedsman Kenn Zimmerman grows spelt, a winter grain relative of wheat. He produces a variety licensed by Purity Foods Inc. for the company to supply its growers.

Wider Rows, Fewer Weeds

Zimmerman quickly found out that the keys to growing organic grain and beans are rotations, cover crops, cultivation and patience. "It just keeps getting better every year you do it, if you stick with it," he observes. "You eliminate what doesn't work and find out what does. You start to get some confidence, and the ground starts to respond to your new practices."

When converting fields to organic production, Zimmerman follows a strict rotation. The last conventional crop in a field is always small grain. After harvest, he seeds hairy vetch, alfalfa or ladino clover to suppress weeds and supply nitrogen for the following corn crop.

In spring, he disks two or three times to kill the cover and flush and kill weeds. Then he field cultivates once before planting corn. "With cover crops, don't expect to have the nice seedbed you'd get with fall plowing. There will be lots more residue. But after disking a few times, almost any planter will work fine," he says. "We're also not as excited about getting into the field as early as everyone else. We really start rolling about the time everyone else is just finishing up planting." That allows Zimmerman to kill more weeds with tillage, and crops germinate faster in the warmer soil to get a jump on remaining weeds.

Zimmerman plants rows east to west to help shade out weeds. Depending on weather conditions, he begins rotary hoeing corn three or four days after planting, and usually hoes three or four times at three- to four-day intervals. Then he cultivates three or four times with an older John Deere solid-shank

Selling cover-crop seed to organic farmers got Zimmerman interested in their practices. Today he sells organic seed stock, doubling or tripling his net compared with conventional crops.

cultivator purchased secondhand for just $100. "I plan to upgrade the cultivator soon," he says. "But it works just fine. We cultivate fast and tight to the plants, and try to throw some soil into the row."

With corn, Zimmerman has returned to wider, 36-inch rows. "The middles are no problem to keep clean. Without herbicides, your troubles are in the row. With wider rows, you have fewer row-feet per acre. So overall, you have fewer weeds," he explains.

Zimmerman says weeds remain his toughest problem. He is considering resorting to herbicides to clean up Canada thistle and morning glory in a field that is still in transition to organic certification.

With no livestock to supply manure, Zimmerman depends on the legume cover to supply all his nitrogen. "We haven't had anything short on N yet," he says. In '90, a very wet season, denitrification and nitrogen deficiency caused problems in many nearby conventional fields. "But our corn stayed green all season. I think it was because of the slow, steady nitrogen release from the vetch. We also got back in the vetch fields a day or two sooner than other fields after it rained," he notes.

> *"People complain about the high cost of seed. But they'll find out what a bargain it is if they crawl around in a combine for three or four hours just to get it clean. There's no easy way to do it."*

Corn On The Cob

High yield isn't necessarily how Zimmerman measures the success of a crop. For example, he grows a corn hybrid selected because it consistently throws large, uniform ears — especially at lower populations and with good fertility. He harvests it on the cob, and the family gathers in the shed on rainy days to grade and bag it as squirrel food for the garden-store market. "So I'm more concerned about marketable ears than overall yield," he says. He also grows open-pollinated corns to be sold for seed — colored varieties for food processing and yellow varieties for livestock feed.

After corn harvest, Zimmerman drills 1 to 1.5 bushels of rye into the stubble without prior tillage, or he broadcasts the same amount and disks it in. The following spring, he usually starts disking the rye when it's approaching full head. "Sometimes it's over the hood of the tractor," he observes. As with vetch covers, he disks two or three times and field cultivates once to kill the rye and weeds.

> *"Most people come to organic farming from the production side. But marketing is everything .... In the seed business, first I learned to sell, then to buy and then to grow. It's the same for organic grains."*

Like his corn, the majority of his beans are in 36-inch rows, and he rotary hoes and cultivates beans on a similar schedule. But in '91, he tried 4 acres of beans in 16-inch skip rows. "I'm testing the tradeoff between having more row-feet with in-the-row weeds and getting a faster canopy. I'm not counting on it, but with narrow rows I may get by with fewer cultivations."

Severe drought burned up the beans in '91, so Zimmerman will have to wait until '92 to find out. "But I'm going to try skip rows again, and this time I'll plant them into a spring-seeded winter grain." Planted that way, winter grains grow only about 6 inches tall, die and leave a weed-suppressing mat of vegetation, he explains.

After bean harvest, he disks stubble once and drills fall grain. He has gradually replaced wheat in the rotation with a licensed variety of spelt that is sold to growers for Purity Foods Inc., a natural-foods processor in Okemos, Mich. The spelt they grow is used in the company's pastas.

Zimmerman grows grains organically, then goes the extra mile to clean and grade seeds such as this spelt for farmers and food processors.

Zimmerman plans to start drilling spelt with 20 to 25 pounds of vetch, then harvest the two together and separate them with his seed cleaner. He hopes enough of the vetch seed will shatter in the field to give him good winter cover and to supply N for the corn to follow.

Zimmerman's goal is to become a regional processor of seed- and food-grade grains and beans for the growing number of organic farmers in Michigan, Ohio and Indiana. "But to expand my capacity, I'll need more capital so I can clean, store, pool and market for these growers."

Zimmerman is optimistic. "Now, the organic foods industry is starting to attract the capital and capitalists needed to develop a solid infrastructure. But until the infrastructure is in place to bring organic foods into the mainstream — including reliable sources of quality seed — they will remain just in the health-food stores."

So far, traditional sources of expansion capital have been scarce, and Zimmerman has had to self-finance his efforts. "You go to the bank and mention organic, and they chuckle and start easing you toward the door," he quips. "They want to know the industry history. And I tell them we *are* the industry, and we're *making* the history."

'The Whole Community Benefits'

Ridge tillage, hard work and a balanced perspective help this farm couple live their vision.

❖

HARLAN, Iowa — Ron and Maria Rosmann don't consider their 480-acre crop and livestock farm typical of American agriculture.

"Our farm flies in the face of all the trends toward fewer and larger farms," says Ron. "People say you can't make it on a few hundred acres. But this farm supports our family plus another full-time person, with virtually no off-farm income."

Rosmann buys few inputs, and relies on low-overhead livestock, good rotations and farm-tuned ridge-till methods that ensure good yields while controlling erosion even on sloping fields. Despite a tight schedule, he squeezes in time for on-farm research as a member and past president of Practical Farmers of Iowa.

He promotes sustainable farming in other ways too. He holds field days to demonstrate energy-conserving, low-input methods, such as on-farm composting and a solar-heated farrowing house. He writes magazine articles and talks with journalists about farming. And he lobbies in Washington, D.C., for alternative agriculture, having served on a USDA advisory committee to promote competitive research grants as part of the "National Research Initiative" authorized by the 1990 Farm Bill.

Ron and Maria Rosmann

Harlan, Iowa

Started cutting inputs: 1983

160 acres corn
100 acres soybeans
50 acres oats
50 acres alfalfa
5 acres rye
67 acres pasture
13 acres tree farm/wildlife preserve
20 acres set aside
15 acres terraces, waterways, buildings

480 acres

Livestock: 60-cow cow-calf herd; 60-sow farrow-to-finish operation

Growing Season: 158 frost-free days; 30 inches of precipitation

Rosmann decided to carry on his family's farming tradition when he graduated from Iowa State University in '73. "I owe a lot to my parents," he says. "Our land would not be what it is today had it not been for their work, care and love. But that's what sustainable agriculture is all about — passing the land on to the next generation in *better* shape than you found it."

For the first 10 years, Rosmann farmed more or less conventionally, using standard fertilizer and herbicide rates, he recalls. He was intrigued by alternative methods, however, including organic farming and ways to cut on-farm energy use.

Looking For Less

Having built a solar-heated farrowing house, the Rosmanns organized and held an alternative-energy fair in the late '70s. "We had representatives from various power companies on hand, including wind-power ones. I even considered putting a windmill on our farm, but I didn't trust wind techology as much as I trusted solar," Ron recalls.

The fair put him in touch with farmers involved with the Center for Rural Affairs in Walthill, Neb. The Center was one of the first grass-roots groups in the Midwest to champion sustainable farming practices. "They sponsored some of the first down-to-earth energy-saving projects for farmers," says Rosmann.

The fair encouraged his interest in large-scale composting, on-farm research and environmental protection. Finally, while on an airline flight in '82, Rosmann read a magazine article about Dick and Sharon Thompson, who were experimenting with sustainable farming practices just 100 miles away.

That summer, the Rosmanns visited the Thompsons, as well as several organic farmers. "I was struck by the fact that many of their fields were cleaner than ours," says Ron.

"Although Dick Thompson had some fairly weedy fields that year, I went home thinking 'This is the guy who will make it work. He's asking the right questions and looking for scientifically sound yet practical solutions.' They were trying to get at the truth, not just promote products, as some organic farmers were doing at that time."

Rosmann spent the winter reading everything he could find about organic farming. "Then we just dove in and 'went organic' on all but the soybean acres in '83." He also joined the Nebraska Organic Growers Association, the predecessor of the Nebraska Sustainable Agriculture Society.

Corn yields dipped, from his normal 125 to 130 bushels per acre to 115, but Rosmann wasn't worried. "I figured it was weather-related, as I had applied enough fertilizer for comparable yields. Instead of applying anhydrous ammonia for 100 to 125 pounds of N per acre as I had been doing in cornfields, I had used an equivalent amount of organic ammonium sulfate. We didn't have yield declines in any other crops."

> *"The nitrate test ... is a good example of how you have to substitute more management if you want to cut chemicals."*

Rosmann figures he *saved* $3,000 on herbicides in '83. "I had excellent weed control with conventional tillage and two or three cultivations. I bought a Buffalo high-residue cultivator in '83, followed by a used 3-point-hitch John Deere rotary hoe in '84. Nature cooperated for the most part our first few years. We had good weather that wasn't overly wet, plus probably some residual effects from previous years."

Corn yields slid further in '84, though, to 100 bushels per acre. "I knew the corn was suffering nitrogen deficiency. But I didn't have enough manure to supply all I needed at less cost than the other organic fertilizers," Rosmann explains.

Adding More N

Besides ammonium sulfate, he also was buying another organically acceptable fertilizer, a Florida-mined product called Cal-Phos. "They were very good fertilizers, but terribly expensive. The cost of product transportation alone is one reason I didn't stay organic in the long run," he observes.

In '83, Rosmann began composting all his hog manure along with straw bedding, year-round. He turns long windrows of material at least twice with a Wildcat compost turner he bought for $7,000 in '83. He spreads finished compost at about 2 to 3 tons per acre in fall or spring, usually ahead of corn.

He composts some of his runnier feedlot manure now, too. Rather than buy bulking material, however, Rosmann still applies some of the feedlot manure directly to selected fields. "I like compost because it improves soil tilth by adding humus, and also kills many of the weed seeds in hog manure," he says. "But it's not a short-term solution to nutrient deficiencies. The nutrients are released slowly, so you have to be patient to see the benefits."

By late '84, Rosmann realized he needed to improve his crop rotation. He had been including oats and forages in the rotation to supply cattle feed, but was growing those crops only on steeper ground. "On the flats and bottomland, I was still just rotating corn and beans. To start solving my nitrogen-deficiency problem, I realized I had to start using longer rotations on that ground too," he recalls.

Today, most of the Rosmann farm is in a five- or six-year rotation of corn-soybeans-corn-oats-alfalfa. Fields that are too steep for soybeans are often in a corn-oats/green manure-corn rotation.

"I also decided to be realistic about nitrogen fertilizer in '85," he recalls. "I wasn't selling any of my grain on the organic market. So I decided to go ahead and sidedress some 28-percent liquid N on corn."

Ridge Till Reduces Weeds

Although two rotary hoeings and two cultivations provided satisfactory weed control, Rosmann was seeing a few more weeds than he could accept. That was especially true with soybeans, which he grew without herbicides in '84 and '85. In fall '84, he started planting rye on bean ground after manuring. The rye cover helped reduce weed levels in the following corn crop. He tandem disked the rye to kill it.

Impressed at how ridge tilling had eliminated foxtail at the Thompson farm, Rosmann rented a ridge-till planter in '85 "just to give it a try." It worked so well that even though a third of his cropland wasn't in row crops, he bought a dealer's demo model as soon as he could, in early '87.

"It seemed the way to go if trying to stay away from herbicides, as I was," he says. And with ridge till, it's hard to avoid investing in equipment, he adds.

"I also wanted to quit compacting soil from disking and quit stirring up weeds in the soil from conventional tillage. Ridge-till equipment would just skim the top few inches of soil on the ridges. It all sounded right," he recalls.

"Of course, we and other farmers in Iowa couldn't duplicate all of Dick Thompson's success in controlling weeds without herbicides, since we have some different soil types, different slopes and so on. We had to make some modifications." Rosmann had his Buffalo planter equipped with disk stabilizers to help keep it on the ridges, even on slopes up to 14 percent.

He was excited about the new direction he was taking. "We had started farming in a holistic manner that fit in with my philosophy spiritually, economically and environmentally," he notes.

He was encouraged by the support he received from the like-minded farmers he was meeting at the Thompsons' annual field days, and at gatherings that led to the formation of Practical Farmers of Iowa (PFI).

Farmer-Scientists

"PFI brought together farmers who weren't growing for an organic premium, but wanted to reduce their inputs mostly for cost and environmental reasons," Rosmann recalls. At PFI's first annual meeting in '86, Dick Thompson was elected president, and Rosmann was chosen as vice president.

Rosmann and many other PFI members adopted Thompson's scientific approach to testing sustainable farming practices. They began conducting rigorous experiments comparing weed-control methods, fertilizer rates and other farming practices.

Starting in '88, for example, many PFI cooperators worked with Dr. Fred Blackmer, an Iowa State University agronomist, to help field test a late-spring soil nitrate test he developed for corn. Using the test when corn is 6 to 12 inches high is one way that Rosmann determines whether or not to sidedress nitrogen at lay-by. In '89, his ridge-tilled corn following soybeans yielded 156 bushels per acre, with just 65 pounds of sidedressed N per acre and no manure or compost added.

"In corn now, I usually sidedress liquid-28 in the range of 50 to 90 pounds of N per acre. I can get by with less if I've added more manure or compost and the weather and temperature are OK. By also considering previous crops and making visual judgments of how the corn looks, I can confidently say I no longer need to use the late-spring test every year.

"The nitrate test isn't foolproof," he notes, based on five years' experience with it. "You still have to think through your overall management when you interpret it. It's a good example of how you have to substitute more management if you want to cut inputs."

Rosmann and others in PFI are continuing to work with ISU to fine-tune the late-spring soil nitrate test. They hope to get a better handle on how manure applications and organic-matter levels influence the cutoff point, the test value above which adding nitrogen provides no economic benefit.

As part of another PFI study, Rosmann started banding 9-18-9 starter fertilizer on some corn acres in '87. The starter helps ensure good early growth even when it's cold on the ridges, PFI cooperators found. The starter also helps the crop get a jump on weeds.

Other than starter and sidedress on corn, Rosmann's only fertilizer purchases since the mid-'80s have been for small amounts of P and K for oats or alfalfa ground. He broadcasts 30 pounds of phosphorus and 50 pounds of potassium per acre, using a dealer blend of diammonium phosphate (18-46-0) or triple superphosphate (0-18-0) and 0-0-60.

Soil tests show most of Rosmann's fields are in the medium to high range for both phosphorus and potassium. Manure and compost applications help maintain P and K levels and are slowly increasing organic-matter levels, Rosmann says. "I hope to continue cutting back on P and K on a field-by-field basis," he notes.

Rosmann still plants some row crops conventionally, partly because of the point rows caused by conservation structures in some of his fields. "I don't like bouncing across point rows," he says. He added buffer strips where he could, and farms strictly on the contour so he can ridge till more acres. "SCS staff estimated my soil loss was at or below 'T' in '91," he notes.

His fields haven't been weed-free, however. Anticipating another drought in '90 following three dry years in a row, Rosmann plowed two alfalfa fields for conventionally planted corn, and cultimulched it to conserve moisture. Then the rains let loose.

The cultimulched ground made a perfect seedbed for grass weeds and was too tough for the rotary hoe to penetrate. "I probably should have banded some herbicide or used a post-emerge spray. But I had used virturally no herbicide in eight years and didn't want to," says Rosmann. His records show he was also saving more than $4,000 annually by avoiding herbicides.

Fine-Tuning Ridge Till

He decided spot treatment of herbicides was in order in '91, though, for two reasons. "It's hard to perfect ridge tillage on sloping ground. And I concluded that I had built a weed seedbank in some of the fields. So I banded some post-emerge herbicides: Basagran and Poast on 40 of the soybean acres at $11.82 per acre, and Accent on some of the corn at $4.05 per acre. I'll wean myself of herbicides again as soon as I can," he says.

Rosmann was seeing more grass weeds than he'd like in his ridges. "I've seen some tremendous weed pressure when I plant soybeans. But I've had courage and still avoid using burndown herbicides. My Buffalo planter and Buffalo cultivator do a phenomenal job of cleaning up broadleaves, but it bothers me to see some foxtail on the ridges.

"We've probably had foxtail seed roll back into the row and germinate. I think we're on the right track by again following Dick Thompson's lead, though."

To minimize soil disturbance on ridgetops, Thompson removed the coulter on his Buffalo planter, split the center depth band and mounted the two sections 16 inches apart. A weed-seed deflector mounted tightly behind the 14-inch sweep, plus residue guards behind the sweep and deflector, ensure weed seeds are moved out of the row area. (See "The Last Chapter Hasn't Been Written," later in this book.)

"We also started drilling covers (rye, vetch and some oats) strictly on the ridges in '91. That's a key: directing covers where you need them. That's a real breakthrough for cover crops," Rosmann believes.

He took the disk stabilizers off his Buffalo equipment in '90, even when planting or cultivating the slopes. "I just follow the mirrors when planting now, and also use a new cultivation guidance system I bought in '90.

"The bottom line on our slopes, however, is that if we can ever no till without any herbicides, I'll probably be one of the first to do it. I can't afford to keep modifying my ridge-till equipment and my combine."

Livestock Can Help

Although they require management time, animals play a key role in the Rosmanns' diversified operation. Cattle and hogs provide low-cost fertilizer and added income. The livestock also have allowed the Rosmanns to hire a full-time employee, who has been a working partner since '77.

"We increased our cow-calf herd from 50 to 60 pairs in spring '92 and are building it to 65. To purchase 80 acres near us for added forage production, we sold to a neighbor 47 more-distant acres we had been renting out. We want to keep most of the 80 acres in pasture and hayfields, and out of row crops," says Ron.

"We've been pasturing cattle successfully in the fall on more than 14 acres of headlands, which include bluestem, Indiangrass and switchgrass. In spring '92, I seeded 13 other acres to native prairie — 6.5 acres of pure Indiangrass and 6.5 acres of pure bluestem." He expects that to be an enjoyable and rewarding project, partly because he's getting serious about improving his pasture use.

"We did some rotational grazing with five paddocks on 40 acres for a few years in the mid-'80s,

Photo Courtesy of Ron Rosmann

A five- or six-year rotation helps Iowa farmer Ron Rosmann harvest crops at or above county-average yields with few off-farm inputs. "Moderate-sized, diversified farms like ours employ more people on the land," he says.

but that just didn't cut it. I was overgrazing the paddocks. I figure rotational grazing will save my existing permanent pasture, which has been less productive with each passing year." He will install more paddocks in '92 and '93 so he can increase the cow herd slightly. That will improve pasture quality and allow better use of pastures.

After years of expanding his hog operation, Rosmann *decreased* his hog numbers from 75 sows to 60 in '92. There were lots of reasons for the cutback, he says. "We stuck to a 52-day farrowing schedule for three years, thinking it was the way to produce hogs. In terms of cash flow, it was. We were cranking out 1,000 to 1,100 head of hogs for sale each year. But we were faced with increasing disease pressure, not enough down time for the buildings, plus *we* needed more rest."

Cash flow became tighter in the mid-'80s for many farmers, including the Rosmanns, who rely on almost no off-farm income. Maria serves as information director for PFI, working at home so she can continue to help rear their three sons, David, 11, Daniel, 9, and Mark, 6.

"We felt we had to raise more hogs to keep the bills paid," says Ron. "By and large, hogs *were* profitable in the '80s. But many farmers increased their hog production to offset grain production losses.

"We cut our crop-production costs dramatically in the mid-'80s by changing our row-crop practices. We had excellent yields with no herbicides until '91, and saved at least $6,000 per year just on fertilizer and herbicides since '83."

To substitute labor for capital in the hog operation, Rosmann built modified Cargill facilities similar to ones the Thompsons had installed. The open-air structures cut construction costs by more than 50 percent, compared with confinement facilities. Rosmann felt that working closely with the animals had its advantages. "We were also expanding the buildings to make things easier down the road.

"But now we're sick of dealing with labor-intensive farrowing facilities, and we're seeing more disease pressure." A pseudorabies epidemic spread to many Iowa hog facilities in '91 and '92, including the Rosmanns' in late '91.

"It's a monkey wrench that could turn out to be a blessing in disguise," he says. "It points out some of the drawbacks of intensive hog production and confinement facilities. I knew the drawbacks of intensive row cropping 10 years ago and should have realized earlier they applied to hogs too. In '92, we went to three groups of 20 sows and let them cycle longer rather than breeding them back immediately after weaning."

Sustaining Communities

The Rosmanns are expanding a small tree nursery they started in '84 as a hobby. They sell a few hundred potted evergreens each year, and plant others on a 13-acre wildlife preserve around the family's pond. The couple anticipate some future timber sales and hope their three sons will maintain the nursery operation.

"I see the tree farm as a future opportunity for the boys. There might not be enough land for each of them to farm, but we hope there will be an opportunity for them to stay here if they choose. Why do so many kids have to leave their home area to find something better?" he wonders.

Rosmann is an active promoter of sustainable agriculture and rural life. "Sustainable agriculture is *more* than a set of farming practices to reduce costs and protect the environment. It includes a vision of the thriving rural communities we want.

"Moderate-sized, diversified farms like ours employ more people on the land, while using less capital and fewer off-farm inputs. That, in turn, has a snowball effect on everything else in the local economy and community. It means more people shopping on Main Street, more kids in the schools and more families in church. That's why we farm the way we do. It's worth it to see the whole community benefit.

"I've never been comfortable with the short-term 'stave off disasters' approach to farming seen on many larger farms," he adds. "The practices we use are more profitable and practical on moderate-sized, diversified farms, like ours. Just one practice won't do it. You have to put them all together in a package.

"We have a total system here: cow-calf herd, farrow-to-finish hogs, forages, small grains, good rotations, even a tree nursery and wildlife preserve. We're striving for a real balance.

"Through our work plus our involvement with PFI, we have the satisfaction of knowing we can influence the direction of agriculture in this country, and improve the quality of rural life.

"Our only regret is we didn't start farming like this sooner."

'I Want To Enjoy My Farming'

Rotational grazing and reduced overhead help this dairyman prosper.

LA CRESCENT, Minn. — You have to do more than just minimize off-farm inputs if you want to be a successful small-scale farmer, says Ken Tschumper. Reducing your capital investment is the next frontier in small-farm survival, Tschumper believes.

"Cutting fertilizers and herbicides are a given when it comes to farming sustainably," he says. "*Capital* costs are what's bankrupting a lot of farmers in the Midwest. I'm making sure my farming lifestyle is a comfortable one, my practices are physically safe and environmentally sound, and my capital investment is less."

As part of that strategy, Tschumper has been scaling up his use of rotational grazing since '88. He had been feeding haylage year-round to his 34-cow registered Brown Swiss herd. Now, he has the animals out grazing six months of the year. To reduce equipment- and maintenance costs even further, he baled most of his alfalfa and oats hay in '92 for winter feed rather than ensiling.

Ken Tschumper
La Crescent, Minn.

Started cutting inputs: 1976
18-20 acres corn
18-20 acres oats
50-54 acres alfalfa/grass hayfields
10 acres pasture
100 acres

Livestock: 34-cow registered Brown Swiss dairy herd
Growing Season: 150 frost-free days; 31 inches of precipitation

Sales of milk, breeding stock and bull calves provide a regular cash flow for Tschumper, who has no off-farm income. "My crops and livestock work together as part of an integrated system. By using sustainable methods, I can make a nice living as a farmer," he says.

Tschumper, 42, is the fourth generation to earn a living from the 100-acre hillside farm, which has been in the family since 1867. An avid reader with a degree in biology and a concern for the environment, he says lowering the farm's use of purchased inputs seemed the obvious thing to do when he took over in the '70s.

Who Needs Herbicides?

After buying the farm from his parents in '76, he began testing ways to cut back on corn herbicides. He saw excellent results by supplementing one or two cultivations with just 1 pound of atrazine and 1 quart of crop oil per acre applied with high-pressure nozzles.

"If you're a cash-grain farmer, it's easy to cut your herbicide use by 50 percent or more just by improving your rotation and fine-tuning your herbicide equipment and timing of sprays," says Tschumper. He considers crop rotation the single most effective weed-control method you can use. He reduces weed pressure by never growing corn in the same field for two years in a row.

"Being a dairy farmer is a real advantage too, since it's easy to incorporate forages in your rotation," he says. Following corn, Tschumper usually plants oats with an alfalfa seeding, harvests oats for hay, then keeps the field in alfalfa for four or five years.

"Once you've cut herbicides by 50 to 70 percent in row crops, it's easy to add rotary hoeing and cultivation to eliminate herbicides. And a cultivation or two can also increase your yields!" he says.

A 1980 visit to a 2,500-acre organic farm in Iowa convinced Tschumper he could eliminate atrazine and rely on mechanical cultivation. "Seeing it done on that large a scale showed me I could do it on my small farm," he says.

Since he was already using an inexpensive front-mount International cultivator, all Tschumper needed was a rotary hoe. He bought a used pull-type IH rotary hoe for $150 at an auction in '82, and converted it from a four-row unit to a two-row. That way, he had some spare parts handy.

Even without herbicides, Tschumper's corn yields stayed essentially the same. "Non-chemical weed control isn't nearly as difficult as some people think.

Photo by Craig Cramer

Planting corn in 16-row contour strips helps Ken Tschumper control erosion on his sloping fields.

A few of my fields were weedier the first couple of years, and I occasionally saw slightly lower yields when I didn't rotary hoe correctly," he recalls. "But it only took me a season or two to learn how to hoe properly. You have to rotary hoe early enough, often enough and with a good enough hoe." Tschumper usually rotary hoes corn twice within two weeks of planting, followed by two cultivations later in the season.

He doesn't agree with folks who say that uncertain weather patterns make mechanical cultivation riskier than using herbicides. "If it's too wet to cultivate, you probably shouldn't be out there spraying, either. There's no reason you can't raise row crops now without herbicides," he adds.

"I have an older pull-type rotary hoe, so I add weight sometimes to penetrate better. It hasn't hurt corn yields. If a couple fields show a little more velvetleaf now, a back-to-back double hoeing handles it."

In 1980, sustainable agriculture was just taking off, Tschumper recalls. "You couldn't just call up your neighbor and ask his opinion about rotary hoeing. There were maybe a dozen farmers in the whole state who had stopped using chemicals. Not being able to draw easily upon the experience of other farmers was the biggest obstacle I encountered when reducing inputs.

"There's more information — and more equipment — available now. We have much better rotary hoes too. There's an economic market for companies to produce improved hoes and guidance-controlled cultivators. Equipment manufacturers are responding to the desires of farmers to reduce chemical use, by providing better-engineered equipment."

Manage Your Manure

Tschumper found that the most important part of his transition to fewer purchased inputs wasn't managing weeds, however. It was managing manure. Until '83, he hauled manure daily. It was a real chore, with spreader breakdowns a common headache. "I was getting tired of fixing it at 20 below," he recalls. He also hated getting stuck in drifted or muddy fields, or burning 2 gallons of fuel just to spread half a load of manure. "I decided it made more sense to stockpile manure."

He began using a concrete cowlot adjacent to the barn as a storage site, with two parallel earthen embankments keeping manure from washing off the lot's lower side. To transport manure, he built a dump wagon for $350 from an old trailer, a used wagon hoist and scrap lumber.

Stockpiling meant no more winter spreading, less wear and tear on equipment and fewer nutrients lost in runoff. Tschumper also fine-tuned his manure applications to use nutrients more efficiently. He used to put most of the manure on corn ground. Now, he spreads at least a thin layer over the entire farm every year, and drags it after hauling to spread it better.

He hauls most manure in spring, and a little in fall. Corn after alfalfa plowdown gets about 10 tons per acre. Corn on lighter soils or following thinner alfalfa stands gets about 15 tons.

Tschumper also spreads manure on hayfields, at 4 tons per acre. "Extension says it's a poor use of nitrogen and will burn alfalfa plants," he cautions. "The trick is to put it on thin and even."

He concedes the alfalfa probably doesn't need the manure N. But the extra phosphorus and potassium help maintain Tschumper's 6-ton-per-acre alfalfa yields. He says alfalfa and manure then provide all the nutrients his corn needs. "The important thing is that, by hauling manure on alfalfa, I can maintain more of the soil's fertility without off-farm inputs. For corn, my plowdown alfalfa easily provides 120 pounds of N per acre, and I don't add any P or K."

Tschumper says his soil-test values are stable. "Even the pH has remained real good. I thought the manure might make it more acidic, but that hasn't been the case," he notes.

Insects have rarely been a problem for him. "We've had alfalfa weevil infestations several times in our area in the past six years. But it's been clear that my farm suffered much less damage than chemical-dependent farms did. We probably have natural predators. I just cut alfalfa as early as possible when there's a weevil scare," he says.

Optimum Productivity

Forage production is the heart of Tschumper's operation. Everything is set up to feed the herd. "Alfalfa is the best crop for these hillsides. It's cheap to grow and is the best feed you can raise for dairy cattle," he says. He seeds alfalfa with an oats nurse crop, and counts on manuring to encourage some grasses to emerge.

Tschumper maintains maximum productivity from his alfalfa by plowing the least productive fields for corn, which he plants in narrow, 16-row contour strips to control erosion. Hayfields are 3 to 5 years old and still contain 50 to 70 percent alfalfa when he turns them under. To get the most nitrogen from alfalfa, he plows it as late as possible — usually within a week or so before planting 95- to 100-day corn hybrids.

"I want to continue to raise some corn. To be truly sustainable, you should raise everything on your farm that you feed your cows, as closely as possible. Buying corn means you're buying into the chemical farming system.

"Because I have a small farm and steep slopes, I raise a smaller amount of corn than most dairy farmers would feed their cows. I compensate by feeding higher-quality forage," he says. "We'd all be better off — even our livestock — if we relied less on grain," he adds.

He spends $100 per ton for a small supplemental ration of ground ear corn and soybean meal with minerals. "That's pretty cheap. The additional protein helps level out any ups and downs in the forage," he says.

Tschumper credits his profitability to many factors in his forage-based system: rotational grazing, Brown Swiss cattle, baling some hay, energy conservation and smart use of part-time help.

Grazing Benefits 'Phenomenal'

During the '88 drought when some of his second-cutting alfalfa was too short to harvest, Tschumper installed temporary polywire fencing. He began rotating the cows through the hayfield during the afternoons. "It's easier for my Brown Swiss, a heat-tolerant breed, to graze in the hot part of the day than it is for them to eat haylage," he notes.

The cows continued to produce well, he says. And fencing costs were minimal — just $110 for a low-impedance charger and a couple hundred dollars for polywire. "I still use homemade wood posts," Tschumper says. "My father had electric fence for 30 years. I've added some Maxishock nine-strand wire and some barb wire. It's quite effective."

He realized early on that you need a good water supply for your livestock. Because his cattle were walking into a couple pasture ponds and pushing in the embankments, he also fenced off the ponds. "Water is the most essential thing to good grazing — you need a good watering system," he says.

In '89, Tschumper divided his 10-acre permanent pasture into several small paddocks. He fed the whole herd haylage in the morning, then had them out grazing the rest of the day, with heifers in a separate paddock.

"A lot of people feel the Brown Swiss is one of the best grazing breeds around. I've never had any other species. My cows like to graze, so I let them do more of that now," he says.

"I do some pasture clipping and manure dragging. But the real key with forage management is using short grazing periods in your rotational grazing. I get a lot of alsike clover and red clover reseeding itself, which won't happen with overgrazing.

"I started reading about rotational grazing in *The New Farm* magazine in the early '80s, and thought it was one of the dumbest things I had heard of. It didn't seem like it would work.

"No one has been more amazed than I — it's almost magical how rotational grazing increases pasture productivity compared with continuous grazing. I went from so-so pasture to one that gives knee-high grass in August after two grazings — that's phenomenal for slopes like mine!"

Tschumper relied on rotational grazing exclusively for six weeks in early summer of '90, when he ran out of haylage. Since then, he's been decreasing the grain ration and use of stored forage during the grazing season, which runs from early May to late November in many years. "Rotational grazing definitely extended the pasture season in '90. I was still grazing dry cows and bred heifers on my permanent pasture on November 29 — that's incredible around here," he says.

With rotational grazing, Tschumper says he needn't worry much about forage-quality testing, either. "The cows tell me it's OK. Whether your grass is 25- or 15-percent protein, you can't pinpoint the quantity eaten, so it's hard to measure protein intake on pasture. But when you have paddocks of nice lush grass, you know it's good quality. Haylage can be visually misleading, as can baled hay."

The benefits of rotational grazing are well-documented now, he adds, and farmers needn't worry about whether it will improve their pasture productivity. "The economics of rotational grazing are so positive. Yet some people put too much emphasis on specific forage-quality figures. It doesn't take too much brains to know you'll get good milk if you put cows out on lush pastures. There are some really sharp Extension forage specialists out there who can help you. But I think too much pasture research is for the benefit of pasture researchers now."

Less Milk, Better Value

Because Brown Swiss typically produce less milk than Holsteins do, Tschumper's rolling herd average of 14,000 pounds is about 2,000 pounds less than the Minnesota average. But his milk value is comparable. Why? "Our milk always tests higher than average in solids," he explains. "Brown Swiss are said to produce the best combination of butterfat and protein. That gives me an advantage,

Photo by Craig Cramer

Tschumper used to feed haylage year-round to his Brown Swiss herd. He and his cows are happier now that he bales more hay and relies on rotational grazing from May to November.

because the growth in dairy sales isn't in fluid milk. It's in cheese and cultured products, where high solids and high quality are worth more."

Tschumper's Brown Swiss average 3.6 to 3.7 percent protein. That brings a 60- to 75-cent premium per hundredweight. Higher butterfat levels, which average 4 percent for his herd, bring a 10-cent premium. In '91, he figures the premiums helped him gross about $3,000 more than he would have for a 14,000-pound milk average from Holsteins.

He has no plans to certify his dairy as organic, though. "I don't think there's enough of a market right now," he says. "If I run short of corn, would I be able to get organically grown corn at a good price?" Trucking costs to an organic cooperative for processing would also be a factor, he observes.

Because Brown Swiss are heat-tolerant, Tschumper taps a lucrative market by selling breeding stock to Central and South American countries. Open heifers fetch $900 and bred heifers $1,200, about $300 more than they do on local markets. He sells 6 to 10 heifers yearly. He also sells up to a dozen bull calves a year to local farmers.

Who Needs A Silo?

Tschumper almost dismantled his silo in '92. Success with rotational grazing, along with low milk prices in '91 and loss of 20 rented acres to CRP, convinced him he could continue to reduce his costs by selling the silo and equipment needed to fill it. He had a buyer lined up for the silo and he baled his first two hay cuttings in '92 before reconsidering.

"I bought the silo new in '77 and it made me money. It was the right decision at that time — ensiling was and still is a good way of storing high-quality forage. But selling the silo seemed the right decision too — rotational grazing allows you to harvest high-quality forage at a much lower cost. And with good balers and timely cutting, you can harvest hay that's *better* quality than silage, at least most of the time.

"It's pretty unusual to take down a silo, and go back to baling hay. I don't know of anyone who has. But the economics of it are so compelling. Without a silo, you can eliminate big tractors, choppers, blowers and forage boxes, save on maintenance and minimize energy requirements — the farm's *and* yours.

"For a small herd, baled hay is the way to go. It takes me just 4 gallons of gas to cut, rake and bale 100 bales of hay — that's pretty cost-effective. For 10,000 bales, that's 400 gallons of gas — that's not much money. Chopping takes twice as much.

"There's more work, cost and danger involved in filling and unloading a silo than in baling. I can hire high school kids to help bale at $5 per hour, and in just a few minutes teach them how to pile bales. Silage help would cost a lot more.

"So I experimented on a grand scale in '92, thinking I could do without the silo. But walking the hillside four to five times daily took more time than I thought it would. The grazing and pasture production were fine, and I like walking, but the time and effort of walking up and down a hill after a day's work took its toll. I'm single, so I'm the one who always has to move the cows.

"I found that rotational grazing is even more farm-specific than I thought. It works great for my young stock, which I don't have to move as often as the milkers. Rotational grazing is practical on most farms but even more so on level ones. On my particular layout, it still takes more time and labor than I'd like."

Lifestyle considerations are important, Tschumper believes. "Farming should be profitable *and* enjoyable, not burdensome. You shouldn't be constantly working," he says. "I want to enjoy my farming and I enjoy living where I do. The country is certainly a nicer place to live than in an urban area. And I have a cleaner farm than many I've seen. I'm making sure my farming practices are environmentally sound.

"The world has changed in the past 15 years. More people are aware of the global environmental crisis. People say American agriculture is the most proficient in terms of productivity, but that's just per number of producers. We actually use much more fossil fuel per unit of product than other farmers around the world do. We have to factor that into the equation of what constitutes good farming.

"I'm trying to position my farm for the next 20 years. I expect to see milk prices below $10 per hundredweight, because the federal government is so greatly in debt that they won't be willing to support commodity prices as in the past. I think free-trade agreements also will likely lower ag prices even further in coming years. So I'm really working at reducing my input costs."

For now, Tschumper figures he'll go back to ensiling his first and third-cut hay, and continue to bale his second cutting. "But I'll keep looking for ways to make my farm very energy-efficient.

Building your farm around low-cost ways of grazing and haying puts you in a better position to buy hay or other feed when the weather works against you.

"My farm continues to be profitable even in hard times, because my input costs are so low. But being profitable isn't a higher priority than environmentally sound farming. The two are mutually supportive. Living with lower prices can best be achieved with a low-input operation, which in most cases *is* an environmentally sound operation."

Encouraging Grassroots Efforts

Tschumper has been a key figure in the "Dump the National Dairy Board Campaign." He and other dairy farmers expect to force a referendum on whether or not the board should be terminated. "We're a grassroots reform movement in the dairy industry, which has been run from the top down by insiders who aren't considering what dairy farmers really want," he says. Many of those farmers are unhappy with the board's use of mandatory checkoffs to promote bovine growth hormone (BGH) and with the board's failure to encourage higher milk prices.

President of the local chapter of the Farmers' Union, Tschumper also participates in a number of other activities that he hopes will help maintain and encourage small-scale farming in the United States. He's a member of the Southeast Minnesota Sustainable Farming Association, a nonprofit group sponsored by the Land Stewardship Project, Lewiston, Minn. He participates in University of Minnesota sustainable agriculture demonstration projects and holds one or two field days annually. Tschumper's confident, outspoken manner and contagious grin make him a popular speaker on the sustainable farming lecture circuit in Minnesota, Wisconsin and Iowa. He also spends a lot of time talking about sustainable farming to schools and civic groups in the three-state area.

"Today, there are really no obstacles to having a reduced-input farm, as far as I'm concerned. It's really not that difficult or risky to eliminate chemicals. The information is all there to do it. I really started putting it all together in '86 and '87. But if I'd known in '81 what I know now, I could have had it together in '83," he says.

"A saying I like to quote is: 'The only safe chemical is one we haven't studied enough.' Inevitably, every ag chemical turns out to be hazardous the more we find out about it. Thousands of farmers are demonstrating — successfully and profitably — that we can farm at least as productively without ag chemicals as we can with them.

Photo by Craig Cramer

Heat-tolerant Brown Swiss are one of the best grazing breeds around, says Tschumper. He taps a lucrative market by selling heifers to farmers in Central and South America.

"The opportunities are there, but we need to get Extension and others out promoting small farms and a more widespread adoption of sustainable practices," he continues. "In just a couple years, we could cut chemical use in half just by getting mainstream farmers to adopt a few practices that we already know will work — like banding herbicides, sidedressing nitrogen, ridge tilling, manure management, rotational grazing and reducing capital costs."

By keeping family farms profitable, such techniques are an effective way to boost sagging rural economies too, he points out. "There can be a lot more enjoyment and a lot less stress in farming — for you and the cows."

Bands, More Beans Cut Chemicals

Smart steel and information slice inputs on MEGA-acres.

❖

DEKALB, Ill. — The advent of an electronic grain flow meter allows the J.P. Faivre Partnership to document in detail its "bang for the buck" on crop inputs. It also takes much of the wiggle room out of chemical recommendations — an important goal of the farm's owners in their information-intensive system.

"There's no more hiding behind test plots. Agronomists will have to figure how to tie inputs to profit or go flip burgers," Steve Faivre says. "We now have whole-field maps in 16-row strips with full data from soil test to planting to yield."

The on-the-go grain monitoring system introduced in August '92 is one of many tools fashioned on the Faivre farm by its spin-off businesses. Each innovation developed as a stepping stone to greater precision in cutting crop-production costs. A long-range commitment by three brothers to reduce inputs led them to put nearly 6,000 acres of crops on ridges, band herbicides and pare synthetic-N use by more than 20 percent since 1985.

J.P. Faivre Partnership

DeKalb, Ill.

Started cutting inputs: 1987

4,338 acres corn
1,617 acres soybeans
0.5 acre wine grapes
5,955.5 acres

Growing Season: 165 frost-free days
35 inches of precipitation

Yet their large scale limits the options open to Steve, who serves as managing partner, and his brothers: Roger, in charge of field operations and cropping; and J.R., who merchandises the farm's grain and manages the partnership's 1 million bushels of commercial grain storage. They've thought through alternatives, but haven't found a way to profitably incorporate cover crops and livestock into their two-crop sequence.

Their plunge into reduced herbicide use on thousands of acres from '87 to '90 — culminating in spot spraying with 10-inch bands and fractional rates — was "too far, too fast. We gave it our best shot. There wasn't much research on how to do it, not on our scale," Steve says.

Weeds that got ahead of the farm staff in those years are still a problem. Band width went up to 12 inches in '91 and to 15 inches the next year.

By '92, Roger found himself expanding his herbicide inventory to include materials for preplant burndown, non-incorporated bands at planting and some post-emergent spot spraying. Herbicide cost was about \$15 per acre for corn and about \$20 per acre for soybeans. The Faivres applied materials at two-thirds to three-fourths of label rate. They ridged all fields in a lay-by pass, and gave some fields an earlier cultivation for weed control.

"Having less ground or more people would have been better, but we tried too many things at once," Roger says. "One year of mistakes takes five to eight years to clean up." He offers this counsel to farmers with large acreage who want to cut herbicides:

❖ **Make sure you understand your weed pressure.** "Keep notes in spring, summer and fall of what weeds are threatening. Match weed presence with each crop, and with those crops in dry and wet, cool and hot years. Watch what weeds follow each crop, too," he says.

❖ **Be prepared to change.** Frost-damage susceptibility in corn residue and weed-control benefits after beans helped the Faivres agree to move up to one-third soybeans in their rotation for '93 and beyond. Areas of high surface residue seemed to suffer worse frost damage than bare ground, they reported.

❖ **Herbicides can mask weed problems.** Weeds you don't see in sprayed areas may be suppressed but not eliminated from the field ecology. Watch for new species to become problems when your weed-control practices change.

❖ **Reduced tillage brings different weeds.** "We'd never had a problem with kochia until '88 — three years into ridge till and the year of the drought. It's still here." Other spray targets in '92 were foxtail, wooly cupgrass, lambsquarters and velvetleaf.

❖ **Take it real slow.** "Take one step at a time. For us, it got tougher as we went," Roger says.

Notching up soybean plantings to about 2,200 acres will push their harvesting capability to the max, but will relieve operator stress at planting time. Storing relatively fewer bushels (soybeans compared with corn) will make more space available in their commercial elevators for outside grain. Other benefits will be to reduce soil-insecticide costs by cutting second-year corn plantings and to increase the amount of corn planted into low-residue, quick-warming soybean ground. Roger says the change also decreases the farm's degree of risk in three areas: mechanical difficulties he's experienced related to ridge tillage into corn stubble; high-residue fields during wet, cool springs; and crop damage from late frosts.

Keeping Tabs

The Faivres work hard to track what happens in their fields as they make changes. To harness electronic data management benefits for crop management, the partnership helped found AgriCAD, an agricultural mapping software developer. With the company's AgMAPP software, operators can combine on-screen an unlimited number of "layers" of information about any field. A color monitor can show weed patches, field boundaries, fencelines, soil types, nutrient variations and application rates for fertilizer, insecticides and herbicides.

In-cab monitors show tractor operators any field history data they call up to help with current operations. Global positioning system (GPS) satellites provide additional precision to AgMAPP users, allowing operators to track their location and follow equipment movement through a field. (Contact: AgriCAD, 7102 W. Fairview, DeKalb IL 60115, (815) 758-8880, fax (815) 756-1007.)

The partnership can pinpoint yields to 16-row strips thanks to the HARVESTYIELD grain flow meter, which is a product introduced by Dawn Equipment Co., another affiliated company. (Contact: Dawn Equipment Co., 1210 E. State St., Sycamore IL 60178, (815) 899-3296, fax (815) 899-3663.) Adding yield information to a field map shows the value of total inputs used in each strip or the whole field in a crop year. Exactly what to make of the complete information set is where Steve Faivre says university researchers should take the lead.

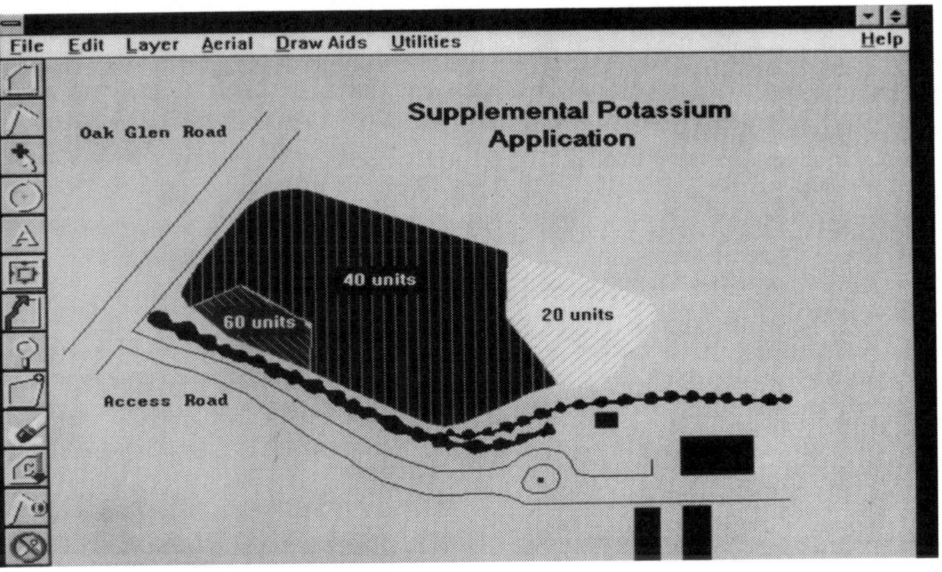

Photo Courtesy of AgriCAD

The Faivre partnership helped launch AgriCAD Inc., a farm software company which markets the AgMAPP field management system. The Faivres use the program in the field and office to show combinations of information "layers" from weed patches to fertilizer application records.

"We developed the technology that ties things together. Now farmers need to learn how to find out the 'Whys?' to make better decisions." Yields are a single measure of a crop's success influenced by a host of conditions during a growing season. Drawing the right conclusion for next year's management is as important as correctly gathering precise statistics on pH, moisture and available N, Steve says.

The basic AgriCAD system, including a GPS receiver, yield monitor, mapping software and radio interface, cost about $7,500 in late '92, $9,000 with computer hardware.

To help interpret how inputs affect yield, the partners are plugging in strip-by-strip plant population statistics gathered electronically at lay-by or harvest. Infrared aerial photos, taken each year to show crop health on the farm and some surrounding comparison fields, also are scanned into the computer maps.

By studying soil types within a single field, the Faivres several years ago began tailoring fertilizer applications to what each section needed. Their custom applicator adapted his rigs to meet their request. The Faivres got fertility only where they needed it and the applicator could offer the same service to other area farmers.

The Faivres' records showed ridge-till corn establishment problems in the late '80s. Causes included uneven seed covering and herbicide effectiveness, residue interference and in-row compaction caused by ridge-cleaning disks. Jim Bassett, a former John Deere engineer, brought his understanding of interlocking toothed wheels to Dawn Equipment. He worked with the partnership to develop and test a zone-tillage tool Dawn calls TrashWheels. Added out front to each planter row, the matched pair of ground-driven steel teeth fling away residue and leave a fluffed soil bed 3 to 6 inches wide.

Benefits include:

❖ Consistent seed depth because of gauge wheel-soil contact, giving better germination and higher populations.

❖ Maximum response from herbicide, which falls on moist, bare soil, rather than on dry residue.

❖ Absence of a compacted layer (from disks or coulters ahead of the drop tube) that may inhibit early downward root development after planting in moist or clayey soils.

❖ Reduction of allelopathic (growth-suppressing) effects of in-row residue on emerging crops.

"Some people get hung up on a specific tillage style, but farmers are moving from different points toward a high-residue, low-tillage system with controlled wheel traffic. That's where TrashWheels fit perfectly," Bassett says.

Steve Faivre hopes zone tillage opens new options for farmers intent on reducing herbicide use. "No till is not the answer. It's a dangerous movement for low-input agriculture. Farmers think it will be quick and easy, but it's costly in chemicals and water quality. They will find that plants need more than water, fertilizer and soil. They need air."

Strong sales for '92 put TrashWheels into service from Michigan to North Carolina on a variety of soils. Bassett says best results came from use on undisturbed soil.

Dawn responded to customer requests for specific tools with several new pro-

(Top) Managing partner Steve Faivre draws upon comprehensive farm statistics to analyze dollar flows in the partnership's farm and commercial grain storage operation. "The bottom line is a compass," Steve says, in his quest to develop ecologically sound ways of managing the nearly 6,000-acre operation.

(Center) Equipment engineer and entrepreneur Jim Bassett heads Dawn Equipment's development of tools for high-residue farming in a variety of tillage systems. He added farmer-friendly features to TrashWheels, the company's first product.

(Bottom) Roger Faivre uses software programs customized by the partnership for its accounting, agronomy and parts inventory needs. Eighteen computer terminals throughout the farm are integrated with computer-assisted-design systems for field mapping and structural engineering.

ducts during '92 (four new patents were pending at mid-year). These include:

❖ Weedwheels row weeder, for farmers wanting to narrow their herbicide bands and bring mechanical weed control closer to their crop roots. TrashWheels are separated to replace barring-off disks, which can prune roots and leave an exposed soil wall to dry out, Bassett says.

❖ A no-till planting variation of TrashWheels that doesn't zone till. Wheels are separated and kept just high enough to fling residue while a coulter set between them does the soil work.

Good Environment

The Faivres' move to ridge till in '85 left more residue on the surface, controlled wheel tracks in fields, reduced spring tillage time and made record-keeping-by-the-pass possible. With the introduction of cultivation and banding, the farm achieved significant herbicide savings. Even with a wider band in '92, weed-killer use was only 60 percent of what it had been in '84.

To further pare costs, the Faivres targeted their largest single fertility input — nitrogen, applied as anhydrous ammonia. To reduce vaporization, they worked with Bassett at Dawn Equipment to develop a better application tool. The resulting Teeknife is an 18-inch sweep that injects ammonia at each wing tip into the root zone of two rows through three-eighths-inch lines.

Since the anhydrous is released 4 inches deep and 9 inches from a vertical escape path, much less of the gas leaks into the atmosphere, dropping the rate required to supply N. "It stays where it does the most good: away from the surface, close to the roots and locked in by moisture," says Bassett. By treating two rows with one knife, Teeknives offer dual benefits of saving horsepower (by pulling half as many injector shanks as every-row models) and allowing hard-to-seal wheel rows to remain undisturbed.

"Our soil health is improving, even with continued use of anhydrous," Roger says. "Earthworms tell me we're on the right track. In some fields, there are so many they're an aggravation when I check seed placement after the planter." Target application rate on corn ground in '91 was 200 pounds of N per acre.

Under study at the Faivre farm is a larger, three-fourths-inch or 1-inch line on Teeknives for application of liquid-suspension fertilizers. The larger line would allow application of the slurry-type material at lay-by.

Dawn Equipment's Teeknife is an 18-inch sweep outfitted to release products at each tip. The tool allows alternate-row operation that seals anhydrous ammonia or high-solid liquid-suspension fertilizer close to the root zone.

As Bassett observed the Teeknives in operation he saw that deflection and differences in wear between row units left the sweeps running at different angles. So he designed a locking pivot-angle adjustment mechanism, which allows operators to set the angle of each knife individually. Level travel in the field means more economical application because less fertilizer is buried below the root zone or able to vaporize from too-shallow placement.

Dismal Drive-Bys

Steering one of the largest farms in DeKalb County from conventional tillage to a high-residue, reduced-spray system was difficult enough for the Faivre partners. It's been even harder for some of their landlords, who haven't liked what they've seen. "It just doesn't have a nice, clean look. It's going to take a generation or two before that attitude passes," Steve says.

He admits there would be less grumbling if the ugly-looking fields consistently produced higher profits. So far, since the partnership's change to ridge till, that hasn't been the case. As the financial specialist of the partnership, Steve can show some of the reasons why.

Farm-generated bar graphs document corn yields above the county average for the past two decades.

The gap is narrowing, however, as others reach the yield potential cap that the Faivres hit earlier through fine-tuned fertility management. What the yield and profit figures *don't* show, however, is that the Faivres have maintained production with fewer inputs. They have slimmed their anhydrous rate by 20 pounds per acre in recent years, and have reduced per-acre costs for capital, fertilizer and other chemical inputs.

"It's like squeezing a balloon. You don't always reduce your costs when you go to conservation tillage or low-input agriculture," Steve says. "You might see significant increased profits down the road, but it's not something that takes place right away."

For the partnership, fertilizer and chemical costs *below* University of Illinois projections are more than offset by *higher* machine and labor costs. Some of the extra cost is for research and development, such as machine testing of Dawn and AgriCAD innovations. The brothers hope to recoup some of that investment with the efficiency gain of improved tools.

Steve places more importance on developing control treatments that rely less on chemicals than on cutting labor expenses. When corn borers arrived in summer '91, Roger Faivre scouted fields 15 times, running computerized cost-benefit projections of possible spray treatments for each check. When the average projected benefit reached $16 per acre (and the maximum benefit hit $37.50 per acre), the brothers chose to spray. They accepted an increased material cost by opting for *Bacillus thuringiensis*, a narrow-spectrum bio-insecticide, to preserve as many beneficial insects as possible.

"We ended up spraying only 150 acres," Steve says. "I heard of one farmer who farms 2,000 acres and sprayed 1,500 acres, simply because he didn't have time to monitor more closely. To me, that's questionable priorities."

Convinced that scouting works, the Faivres hired a pest-management firm to walk their fields in '92. The move brought professional expertise more directly to their fields, linked the farm with scouting reports from other insect-monitoring traps in the area and gave Roger more management time for other cropping decisions.

Go For Seed Details

Part of Roger's job is selecting the most profitable seed varieties. "We pay attention to any information seed companies have on how varieties perform on different soil types and after other crops. I work with 40 soil types, and yields can vary up to 100 bushels per acre with the same variety.

"In fields that are the most variable in soils, I go with more versatile varieties, even though they may not be the top yielders if conditions are ideal," Roger explains. He's learned it pays to be direct with seed sellers. "If you don't ask detailed questions, you get a generic answer. Ask, 'In order for your company to perform well for us, what are the best and worst conditions for variety X and variety Y?' They'll begin to know what kind of answers you want. Some tell me, 'Don't put this one after last year's corn.'"

Attention to details and a non-evangelical approach to change have earned the Faivre farm grudging respect in DeKalb County, according to Lyle Paul, superintendent at the Northern Illinois Agronomy Research Center. As the former county Extension agent, he knows a single family can't rent 6,000 acres without causing some hard feelings. Despite that now-diminishing background sentiment, Paul says the Faivres are viewed as farmers who aggressively pursue sustainable farming changes in a realistic manner.

Steve Faivre may *never* be satisfied that he or those he works with have done enough to be truly sustainable. After spending eight intense years developing new ideas, however, he says the management, agronomy, software and machine-technology parts of the operation are starting to show benefits. While each part holds economic promise, it is their *combination* that he believes may eventually provide an economic edge for large-scale sustainable systems.

Chapter 3: When In Doubt, Ask A Farmer

'FONE' network helps show the way to sustainable production.

EMMAUS, Pa. — Nothing beats the advice and experience of other farmers, as the previous chapters show. A lot of the questions and issues these farmers faced were ones *The New Farm* editors have grappled with for years: What *really* happens to a farm's income when chemicals are reduced or eliminated? What happens to yields? Are weeds, nutrients or other factors as much of a problem as many think?

Starting in early '84, we posed these and other questions to the largest group of farmers we know of who are cutting back on chemical use: readers of *The New Farm*. A questionnaire for this purpose first appeared in the March/April '84 issue of the magazine. Through 1987, *The New Farm* regularly carried it or a similar form that readers were asked to fill out.

Respondents automatically became members of "FONE" — the Farmers' Own Network for Education. "The purpose was to identify farmers who were implementing innovative farming practices. Their advice could serve as a tool for those at the Extension and university level, but most importantly, to link farmers with other farmers," notes Anne Schauer, former FONE information coordinator.

By March '85, more than 700 readers from 43 states and seven Canadian provinces — and even a few readers in Europe — had joined FONE. They told us *why* they wanted to use fewer chemicals, *what* obstacles they encountered when they began doing so and *how* they overcame them.

With support from the Regenerative Agriculture Association (a predecessor of the Rodale Institute and then-publisher of *The New Farm*) and a grant from the Rockefeller Brothers Fund, we compiled and mailed to FONE participants in spring '85 a "Summary Report on Farming Operations" of FONE field-crop farmers. Many of the report's composite figures are worth noting.

Farm sizes range from just a few tillable acres to more than 3,500. The U.S. responses average 306 acres, compared with 498 acres in Canada. About half the farms are from 1 to 200 acres, 28 percent are smaller than 100 acres, and 25 percent are larger than 400, with 9 percent exceeding 750 acres.

Corn is grown on 66 percent of the farms, soybeans on 42 percent, wheat on 45 percent and hay on 32 percent. Hay and small grains are grown on 55 percent of the farms. Other cash crops include: straw (21 percent of the farms), fruits and vegetables (16 percent) and assorted specialty crops such as dried beans, seed crops and cotton (fewer than 5 percent).

More than 80 percent of the farmers raise at least one type of livestock. Almost 60 percent raise beef cattle, 35 percent raise hogs, 33 percent raise dairy cows, 25 percent raise poultry and 15 percent raise sheep. Horses, goats and other livestock are raised by fewer than 14 percent of the farmers.

Fifty-three percent of the farmers have been cutting purchased inputs for three or more years, starting in the late '70s and early '80s. Only a handful are long-time organic farmers.

'Cheaper, Safer, Healthier'

"We do this because it is cheaper, but also because it is safer," wrote Richard J. Teubel, a 185-acre, cash-grain and dairy farmer in Bloomfield, Iowa. Teubel's remarks describe the feelings of many of the other farmers we surveyed. Many gave multiple reasons for cutting back on chemicals, but by far the most popular were "Cut production costs" (82 percent), "Environmental concerns" (76 percent) and "Personal/family health and safety" (73 percent). Fewer than 18 percent expected to receive market premiums for chemical-free products.

Many farmers offered practical, although fairly general, advice. "Fit your system to your climate and soil," said Texan Mackie Allgood, a cattle, sheep and small-grain producer. "Be flexible. Enjoy farming enough to *work*, or do something else. All good farming is labor-intensive."

"You should enjoy yourself and what you're doing," agreed Norm Zeman. "It's great. It's a way of life." Zeman, one of the few long-time organic farmers who answered the questionnaire, raises beef cattle and hogs, and rotates corn, beans, oats and hay on 120 acres in Ottosen, Iowa. "There is no right way to farm," he noted. "Do whatever works for you. But don't expect purchased products to perform miracles." And finally, "[Reducing chemicals] won't be easy. You might be talked about!"

Judging from our survey results, though, the only gossip these farmers have had to endure was pretty favorable. Nearly two-thirds of them said crop yields either stayed the same or increased when they started using fewer chemicals, while just 35 percent noticed yield declines for some crops.

Eighty-three percent said their net income either remained the same or increased after they reduced off-farm inputs.

"This is just the second year we have partially farmed organically," said Pennsylvanian Robert Smith, a beef and dairy producer farming 468 acres. "Corn after clover/alfalfa received no fertilizer; 10 tons of manure per acre was applied. We are still using a herbicide — no insecticide, though — but would like to change to cultivating in future years. Yields were about the same on this land as on other corn ground where chemical fertilizer was used according to soil test."

The farmers who did report yield declines generally felt it took from two to three years for them to recover, a conclusion shared by scientists at the Rodale Institute Research Center and at other research facilities that have studied what happens when chemicals are withdrawn from a cropping system. A few survey farmers said yields rebounded in just one year, while others said yields still hadn't recovered.

Using On-Farm Resources

What changes had FONE farmers been making? Most had reduced or quit buying fertilizers (72 percent) or pesticides (61 percent). Fifty-nine percent were still using some synthetic fertilizer, however, while an almost equal number (54 percent) relied on fresh animal manure to supply at least a portion of their crop nutrient needs. Organic fertilizers and composted animal manures were each in use on about one-fourth of the farms, and sludge and other organic by-products on fewer than 8 percent.

More than half of the farmers had started using green manures and cover crops (55 percent) and better rotations (51 percent). Of those using green manures and cover crops, farmers preferred clovers (56 percent) and alfalfa (50 percent). Small grains are used by 39 percent of those who reported using green manures and covers.

Did farm-management changes such as these affect net farm income? You bet they did! The effect was even more pronounced than the reported yield changes. A whopping 83 percent of the 447 farmers who responded to this item on the questionnaire said their net income either stayed the same or *increased* when they began farming with fewer purchased inputs. Just 15 percent said net income declined.

Surprisingly, having livestock didn't seem to offer any particular advantage to yields or income. When compared with the respondents as a whole, slightly more of the farmers with livestock said they noticed yield declines when they began reducing chemicals. But slightly more of them noticed yield *increases*, too. Likewise, *income* increases and declines were reported by an almost identical percentage of respondents, both when livestock farmers were included in the total, and when they were examined separately.

R.E. Corcoran of Elvins, Mo., credited a flexible crop plan with maintaining cash flow on his 140-acre, mixed crop and livestock farm. "For several years, inputs were reasonable and I did well. But I have always kept track of *costs to plant.* And as prices changed, it became obvious that it would be profitable to go back to grass, and buy corn again.

"In this area, hay is more profitable, provided you have the equipment and some available laborers," explained Corcoran.

"I've found out that if you plant rye in cornfields after harvest it helps control weeds and grasses," wrote Illinois dairyman David L. Marcure, who says both his yields and net income improved when he cut back on chemicals. "I also plant rye in the fall in cornfields that are going to be in oats the next year. Then, in the spring, I just disk and harrow (before planting). There will still be some rye ... but not enough to hurt anything. You have a much cleaner stand of oats with hardly any weeds or grass."

John Myer, a mixed-livestock producer who farms 300 acres in Ovid, N.Y., thinks herbicides simply aren't cost-effective in small-grain and hay crops. "I haven't used herbicides on small grains or alfalfa seedlings in three years, with *no* reduction in yield. The key is to remain flexible in your crop rotation. Plant a crop that will do well in that field *that* year. If a field is too wet to work early, plant a warm-weather crop."

Cash-grain farmer Larry Olson, who raises hay and mixed grains on 180 acres in Granite Falls, Minn., seemed to take a page directly from Dick and Sharon

Thompson's weed-management textbook: "The less tillage I did, the less problem I had with weeds. That's why I've moved to ridge tillage," Olson wrote.

The Thompsons, contributing editors to *The New Farm* during much of the '80s, made a similar conclusion about tillage in the early '80s when they began experimenting with ridge-planted corn and soybeans on their 300-acre Iowa farm. "We have found that the *deeper* you till and the *more* you till, the *more weeds* you'll have," say the Thompsons.

Illinois farmer Mike Raftery, a 120-acre, hog- and beef farmer, says he also started ridge tilling as part of his low-input program. "I feel it gives me all the benefits of a minimum-till situation and the advantage of good mechanical weed control." Yields of corn, soybeans, small grains and alfalfa hay have stayed the same on Raftery's farm, while his net income has increased. What's more, he added, "the tilth of the soil is really coming back."

Not all of the farmers in our survey had to invest in new equipment. More than a third said they bought no new equipment at all when they decided to cut back on chemicals. Of those who said they did have to visit their machinery dealers, 20 percent bought cultivators, 14 percent bought chisel plows, 12 percent obtained rotary hoes, 6 percent bought manure spreaders and 5 percent bought haying equipment.

Obviously, mechanical weed control was high on the list of priorities, though the means for achieving it varied widely. For example, Richard Laska, of Winona, Minn., said he swears by his rotary hoe. "A rotary hoe is a must if you don't use herbicide," he wrote. Yet David D. McCoy, who raises beef and dairy cattle on 360 acres in Ohio, sounded like he was about ready to sell his. "I need to find a better way to control weeds *in the rows* of my corn. My rotary hoe isn't doing a good job."

Perhaps Texan Rick Watzl, who began reducing chemicals on his 700-acre cotton and grain-sorghum farm in the early '80s, best captured the respondents' feelings: "Apply herbicides only when needed and not as a habit."

Three-fourths of the farmers quit using chemicals for financial, health or environmental reasons — usually all three.

Of those farmers who saw yield declines when they cut back on chemicals, many offered a reason — usually several reasons. Weeds caused yield reductions for 156 of the 163 farmers who reported yield declines. That's 25 percent of all survey respondents.

Nutrient deficiencies reduced production on 109 (17 percent) of the farms. Almost a third of the survey respondents reported that nitrogen was their most deficient nutrient when cutting back on chemicals. That's probably why so many of the respondents include leguminous cover crops in their rotation. Insufficient phosphorus was cited by 17 percent and low potassium levels by 12 percent. Lime, sulfur, magnesium or other nutrients were mentioned by fewer than 6 percent.

A few farmers said poor weather, lack of proper equipment or a lack of know-how were their main obstacles to cutting purchased inputs. Only 2.5 percent of the farmers — 16 in all — said insects were responsible for lower yields when chemicals were withdrawn. Only five cited plant diseases.

That weeds were the most frequently mentioned problem probably comes as no surprise. But many of these farmers succeeded in controlling weeds with less or no herbicide, and reported higher net incomes now that they're relying less on chemicals for weed control.

Tapping Farmers' Experience

FONE columns in *The New Farm* during the mid-'80s regularly featured field- or farm-proven tips provided by network members to help other farmers switch to fewer off-farm inputs. By January '86, the FONE network had expanded to almost a thousand strong. The FONE database allowed *New Farm* editors to access farmers for advice about region-specific methods and technique adaptations. To address reader questions about using legumes in low-rainfall areas, for example, the database identified 80 farmers in semiarid regions who were successfully using legumes to cut costs and control erosion. We called many of them and included their advice in a *New Farm* article packed with tips. (See "Home-grown N — Dryland Style," *The New Farm*, January '86.)

In succeeding issues, FONE columns in *The New Farm* often focused on a single farm topic or production practice. In the February '86 issue, for example, Nebraskan Rich Mazour described how he saved $18 per acre on fertilizer for wheat by banding starter P and taking deep soil samples for residual N.

The March/April '86 FONE column told how

Canadian grain farmer Louis Riou relied on hardy varieties and mechanical cultivation to eliminate more than $30 per acre in herbicides. One of his tricks was to aggressively till Canada thistle, his toughest weed, during summer fallow to starve root reserves.

Ohio dairyman Dave McCoy, a frequent FONE-tip contributor, related how he dropped 80 rental acres to focus more on rotational grazing of his 35 acres of grass/legume pasture. Five 7-acre paddocks reduced his haying costs by half and allowed him to feed more cows at no additional cost. (See "Bucking The Dairy Buyout," *TNF*, September/October '86.)

In a later issue, McCoy described how he made what he called a "spider harrow" by replacing the blades on his John Deere finish disk with 32 spiders. The rig helped McCoy reduce tillage without using chemicals. (See "From Residue To Seedbed In One Pass," *TNF*, February '87.)

Farmer surveys continued to provide the FONE network with information in the late '80s. Rodale Institute staff presented advice and information assembled from the FONE databank on many occasions to USDA, state departments of agriculture, research and Extension, and congressional officials.

Low-Input Trends Continue

With the help of a grant from the Charles Stewart Mott Foundation, we sent a "special farmer survey" to 8,000 *New Farm* subscribers in Ohio, Illinois, Indiana and Michigan in early '89. Survey questions focused on the changes in field practices that farmers in these states were making as they moved toward more environmentally sound farming.

As with previous surveys, the intent was to show how reduced reliance on off-farm inputs affected the overall operation of farms, and to document farmers' reasons for making the changes they had. Although not necessarily a cross-section of American agriculture, the survey results could be compared with other surveys over time to indicate important trends in American farming. (See Chapter Four, "Forget Conventional Wisdom.")

Of the surveys sent, 843 were returned. They show that respondents cut chemicals mainly for financial, environmental and personal or family-health concerns.

Following on the heels of high debt loads carried by many farmers during the mid-'80s farm crisis, that's not surprising. The farmers want their operations to be more profitable. The figures could also imply synthetic fertilizers and pesticides are just too expensive, or farmers are using more than optimum amounts.

"Money spent on corn herbicide is a complete waste," an Illinois dairyman wrote on his survey. "Most herbicides are incorporated with a [field- or row-crop] cultivator anyway, so why not just cultivate twice more with ridging equipment just to cover 'in row' weeds and pocket the herbicide money?"

"Herbicide costs can be cut by three-fourths by banding and cultivating," an Ohio cash-grain farmer commented. "I'm not giving all my money to some chemical company and I'm not polluting everything," he added.

Most respondents (71 percent) expressed concern for the environment as a reason for reducing chemicals. Media coverage of environmental issues had increased in the late '80s, compared with earlier years. Mainstream journalists as well as agricultural reporters frequently pointed out examples of poor use of natural resources, from bird kills to chemical spills to crowded landfills. Even in our earlier surveys, though, close to three-fourths of farmer-respondents cite environmental safety as a major reason for cutting chemicals. Regardless of how they describe their farming system, most farmers really care about the condition of their land, *New Farm* editors have found.

Many farmers don't like the health risks associated with the use of agrichemicals, however. Sixty-three percent indicate personal or family health was a major reason for cutting chemicals.

Somewhat surprising: Fewer than 17 percent said market premiums were a factor in their decision to cut chemicals. Organic and other specialty products might bring higher returns, but most farmers aren't out to make a killing in market niches.

How did cutting agrichemicals affect the respondents' farm operations? In the tables that follow, we present some of the key results of this survey. The tables display the actual number of responses for that particular choice or answer as well as the percentage of respondents who listed that choice.

Almost 94 percent of the farmers who returned a questionnaire indicated they would be interested in the survey results. That shows the strong interest farmers have in learning more about what other farmers are doing to reduce their use of purchased inputs.

"Since FONE's beginning, a number of farmers have willingly shared information about themselves and their farming careers," notes Anne Schauer, former FONE information coordinator. "Without this interest and commitment of the FONE participants, the movement toward a more environmentally sound agriculture would not be as widely accepted as it is today."

Breakdown of Acreage in FONE

Although respondents farm a wide range of acreage, almost 70 percent of them farm 400 acres or less.

# Responded	814
0-50 acres	86 (10.56%)
51-100 acres	124 (15.23%)
101-200 acres	164 (20.14%)
201-300 acres	105 (12.89%)
301-400 acres	90 (11.05%)
401-500 acres	63 (7.73%)
501-600 acres	40 (4.91%)
601-700 acres	35 (4.29%)
701-800 acres	24 (2.94%)
801-900 acres	19 (2.33%)
901-1000 acres	20 (2.45%)
1000 plus acres	44 (5.4%)

Type of Enterprise

Cash grain (63.9 percent) and livestock (14.6 percent) are the most common enterprises found in this survey. Only 73 dairy farmers returned the survey, or 9.0 percent of respondents, and only 28 farmers (3.4 percent) indicated they market produce as their primary enterprise. So this is mainly a survey of cash-grain and livestock farmers. Because 225 of the cash-grain farmers also raise livestock, 15 of the dairy farmers also produce cash grain and 11 of the produce growers also raise grain, the results presented here were not adjusted to represent just cash-grain farmers or livestock producers.

# Responded	814
Cash Grain	520 (63.88%)
Livestock	119 (14.61%)
Dairy	73 (8.96%)
Mixed Enterprise	43 (5.28%)
Market Produce	28 (3.43%)
Hay	7 (0.85%)
Woodlot	4 (0.49%)

Livestock Raised

Beef and swine topped the list for the 78.7 percent of respondents who raise livestock. Fewer than 15 percent had dairy cows.

# Responded	718
Beef Cattle	306 (42.62%)
Hogs	189 (26.32%)
Dairy	105 (14.62%)
Sheep	75 (10.45%)
Poultry	48 (6.69%)
Horses	31 (4.32%)
Goats	25 (3.48%)
Do Not Raise	153 (21.31%)

Major Crops Produced (Grains)

Of grains, respondents raise primarily corn (85.5 percent), soybeans (69.3 percent), wheat (55.5 percent) and oats (29.2 percent).

# Responded	805
Corn	688 (85.46%)
Soybeans	588 (69.31%)
Wheat	477 (55.52%)
Barley	26 (3.22%)
Oats	235 (29.19%)
Rye	25 (3.1%)
Milo	11 (1.36%)

"The original idea behind FONE was to link farmers as a means of exchanging ideas and information. Over the years, FONE ... served not only farmers, but researchers and agricultural policymakers as well."
— Anne Schauer

Major Crops Produced (Legumes/Hay)

Of legumes and hay, a whopping 85.5 percent of respondents raise clover. Most raise it for hay and pasture, but more than 40 percent report that they also use it for a cover crop and green manure. Only 22 percent grow alfalfa.

# Responded	805
Clover	688 (85.46%)
Pasture	291 (36.14%)
Mixed Hay	279 (34.65%)
Alfalfa	177 (21.98%)
Oilseed	13 (1.66%)
Dry Beans	9 (1.11%)

Soil Conservation Practices Used

Most respondents (82.9 percent) rely on crop rotation as the primary technique for conserving soil. "I have found crop rotation to be a must," wrote one Ohio farmer, who says his neighbors couldn't believe it when he harvested 48-bushel wheat yields in a drought year — without any synthetic inputs for two years. Cover crops, grass waterways and conservation tillage are also used by about half of respondents.

# Responded	788
Crop Rotations	653 (82.86%)
Cover Crops	421 (53.42%)
Grass Waterways	390 (49.49%)
Conservation Tillage	377 (47.84%)
Contour Cropping	90 (11.42%)
Strip Cropping	86 (10.91%)
Terracing	66 (8.37%)
Mulch/Crop Residue	58 (7.36%)
No Till	32 (4.06%)
Ridge Till	24 (3.04%)
Other	2 (0.25%)

Primary Tillage

Moldboard plow, chisel plow and offset disk are the major primary-tillage devices reported. About 16 percent of the farmers indicate they were using no tillage on some of their fields. Fewer than 7 percent of respondents were using ridge-tillage equipment.

# Responded	801
Moldboard Plow	360 (44.69%)
Chisel Plow	343 (42.83%)
Offset Disk	271 (33.83%)
Ridge-Tiller	52 (6.49%)
Subsoiler	38 (4.74%)
Tandem Disk	21 (2.62%)
Rotavator	18 (2.24%)
Cultivator	17 (2.12%)
Harrow	6 (0.74%)
Ripper	3 (0.37%)
Other	17 (2.12%)

Primary Mechanical Weed-Control Methods

Almost 80 percent of respondents use a cultivator, with the larger the acreage, the greater the reliance on a row-crop cultivator. Fewer than a third rotary hoe, but 37 percent of farmers with more than 400 acres rely on it, vs. 29 percent of smaller farmers.

"The wonderful improvement in the texture and quality of the ground after I started farming on ridges was amazing," wrote an Indiana dairy farmer. "It's so simple and saves so much fuel oil, soil and toil."

# Responded	789
Cultivator	629 (79.72%)
Rotary Hoe	250 (31.68%)
Mower	217 (27.5%)
Mulch	38 (4.81%)
Propane Torch	4 (0.5%)
Fallow	3 (0.38%)
Sheep/Goats	1 (0.12%)
Other	17 (2.15%)

Green Manures and Cover Crops

Almost 80 percent use green manures and cover crops. Favored are clover, alfalfa, rye and small grains. "In the last two years we have begun using rotations and green manures extensively," wrote an Illinois grain and livestock farmer with more than 400 acres. "Concurrently, we are experiencing our third year of extreme drought. This has made the green manures less productive than hoped for, especially the vetch, but has impressed upon us the risk-spreading benefits of a diverse crop mix. We are excited about the future."

# Responded	783
Clover	315 (40.22%)
Alfalfa	284 (36.27%)
Rye	279 (35.63%)
Small Grains	272 (34.73%)
Grasses (Mixed)	117 (14.94%)
Vetch	40 (5.1%)
Buckwheat	13 (1.66%)
Brassicas	2 (0.25%)
Other	21 (2.68%)
Don't Use Cover Crops	159 (20.3%)

Participation in Government Programs

About two-thirds of respondents participate in a government program, mostly for commodity support and set-aside. Fewer than 3 percent were enrolled in CRP, however.

"I believe if all government programs were cut out more farmers would do a better job of farming. The government programs that are made for the big farmers will, and do, put the small farmer out of business. Being subsidized does not help the farmer look for better ways of farming, like low-input," said an Indiana grain farmer with less than 200 acres.

"I'm trying to get back to a corn-soybeans-wheat rotation," said a 350-acre Illinois farmer. "Government programs made it impossible to maintain that rotation, but this is the last year I'll be in the commodity support program."

# Responded	755
Commodity Support	239 (31.56%)
Set Aside	236 (31.25%)
Conservation Reserve	21 (2.78%)
All Three Programs	42 (5.56%)
Do Not Participate	254 (33.64%)

Nutrient Sources

Synthetic fertilizer topped the list, with more than 72 percent of farmers buying it. Almost half report using fresh animal manure, while more than a third use green manures and cover crops. About one-fourth of the farmers use rock sources of P and K, while fewer than 20 percent use other "natural" or organic products. Almost 15 percent compost manure, although that figure could include aging or stockpiling of manure rather than complete composting.

# Responded	805
Synthetic Fertilizers	583 (72.42%)
Fresh Animal Manure	372 (46.21%)
Cover Crops	300 (37.26%)
Natural Rock P&K	185 (22.89%)
Composted Animal Manure	117 (14.53%)
Natural Products	48 (5.96%)
Poultry Litter	39 (4.84%)
Kelp Products	32 (3.97%)
Sludges	28 (3.47%)
Fish Emulsion	26 (3.22%)
Organic Fertilizers	9 (1.11%)
Other	9 (1.11%)

FONE Encouraged On-Farm Research

Rodale Institute's national networks for on-farm research began in the mid-'80s as an extension of the FONE database. Many of the farmer-cooperators were identified through FONE surveys and were featured in *The New Farm* magazine.

The farms represent may different soil types and climatic conditions. Cooperators conduct research geared to answering questions and problems specific to their region. The research plots, as well as whole farms, serve as demonstration sites for other farmers, researchers and agriculture professionals. Annual field days provide a meeting place for idea exchange. The farmer is put in the position of instructor in an informal hands-on classroom.

For more information about Rodale Institute's national networks and farm tours, write: Field Days, Rodale Institute, 222 Main St., Emmaus, PA 18098.

Farming System

Almost 40 percent describe their farming operation as conventional. A third say they are "low-input," while almost 18 percent are "sustainable." About 9 percent use the term "regenerative" and about 9 percent say they are "organic." Of organic respondents, only a third are certified or in transition.

# Responded	755
Conventional	296 (39.2%)
Low Input	250 (33.11%)
Sustainable	134 (17.74%)
Regenerative	71 (9.4%)
Organic *	70 (9.27%)
Natural	58 (7.68%)
Ridge or No Till	9 (1.19%)
Conservation	3 (0.39%)
Other	32 (4.23%)
*** Certified Organic**	
# Responded	**65**
Certified	10
In Process	14
Not Certified	41

Changes Implemented

In cutting inputs, more than half of respondents say they have reduced their use of fertilizer, herbicides, insecticides and tillage. One of every four has started using cover crops and a similar percentage have started a crop rotation.

Most farmers tend to implement more than a single change at a time, the survey shows. Where more than one change is made, use of green manure and/or crop rotation are usually added, especially where fertilizer and herbicide reductions are made.

# Responded	686
Reduced Fertilizer	383 (55.83%)
Reduced Herbicide	362 (52.76%)
Reduced Insecticides	355 (51.74%)
Reduced Tillage	352 (51.31%)
Started Using Cover Crops	173 (25.21%)
Started A Crop Rotation	163 (23.76%)
Started Using Green Manures	138 (20.11%)
Increased Tillage	39 (5.68%)

How Long Since Changes Started?

More than half (58 percent) had been cutting off-farm inputs for more than three years. That gives strength to many of the survey findings, since they're based on farmer experience with input reductions rather than wishful thinking. Only 20 percent of respondents had not reduced inputs when this survey was taken.

# Responded	762
More Than 5 Years	226 (29.65%)
3 to 5 Years	216 (28.34%)
2 Years or Less	142 (18.63%)
Haven't Started To Reduce Inorganic Inputs	155 (20.34%)

How Reduced Inputs Affect Yields

Don't expect results overnight when cutting your inputs. But don't worry too much about yield declines, either. While 7 percent of respondents say they saw yield *increases* right away, almost 62 percent of those who reduced or quit inputs report *no change in yields.* Only 30 percent saw yield reductions.

# Responded	568
No Change In Yields	350 (61.61%)
Yields Increased	42 (7.39%)
Yields Decreased	170 (29.92%)
Increased & Decreased	5 (0.88%)
Some Decreased, Some No Change	24 (4.22%)

Years Until Yields Returned

Of 170 farmers reporting yield declines, slightly more than half (95 farmers) responded to the question asking if their yields had returned to the levels seen prior to input reduction. A third of these were still waiting for yields to return to previous levels. Don't let "a third" scare you. It represents only 31 farmers waiting for yields to recover after reducing off-farm inputs — that's less than 4 percent of all farmers who returned a survey. Another third (another 4 percent of all survey respondents) weren't sure if their yields were back to their previous levels yet. That says that 92 percent of those who cut chemicals saw *no* yield losses or *no long-lasting yield losses*. Obviously, though, factors other than just chemical use affect yields.

# Responded	95
1 Year	5 (5.26%)
2 Years	9 (9.47%)
3 Years	10 (10.52%)
4 Years	4 (4.21%)
Have Not Returned	31 (32.63%)
Uncertain	31 (32.63%)

Reasons for Yield Decline

Taking the drought of '88 into mind, one might have expected weather to be the most important factor cited by farmers reporting decreased yields. Unfortunately, "weather" was not specifically listed as a choice for this question on the survey form, and only 28 of the 160 farmers who responded wrote in "weather" or "drought" as a factor. So it's hard to accurately summarize the survey results for this question. Most respondents (118 of them) checked off "weeds" as a cause of crop-yield declines, followed by nutrient deficiency (79 responses). Only 19 reported having insect problems hurt yields.

# Responded	160
Weeds	118 (73.75%)
Nutrient Deficiency	79 (49.37%)
Weather	28 (17.5%)
Insect Problems	19 (11.87%)
Equipment Problems	19 (11.87%)
Plant Diseases	7 (4.37%)
Chemical Residue	1 (0.62%)
Other	2 (1.25%)

Labor Needs Affected

About half of respondents report that their labor needs increased (40 percent said slightly; 11 percent said significantly) after cutting inputs. The other half saw no change or even a decrease in labor needs.

# Responded	528
Remained the Same	150 (28.4%)
Increased Slightly	212 (40.15%)
Increased	59 (11.17%)
Decreased Slightly	80 (15.15%)
Decreased	27 (5.11%)

Equipment Changes Needed

Almost 60 percent of respondents who made changes in their farm practices say the changes did not require purchase of additional equipment. The need to buy mechanical weed-control equipment became apparent to 101 respondents — 65 farmers needed cultivators, 36 bought rotary hoes.

# Responded	517
No Change	309 (59.76%)
Modifications Only	40 (7.73%)
Ridge-Till Equipment	35 (6.76%)
No-Till Equipment	36 (6.96%)
Rotary Hoe	36 (6.96%)
Cultivating Equipment	65 (12.57%)
Primary Tillage Equipment	21 (4.06%)
Secondary Tillage Equipment	8 (1.54%)
Subsoiling Equipment	5 (.96%)
Finishing Equipment	3 (.58%)
Haying Equipment	10 (1.93%)
Harvesting Equipment	5 (0.96%)
Material Handling Equipment	20 (3.86%)
Planting Equipment	11 (2.12%)
Miscellaneous Equipment	9 (1.74%)

Effect on Net Income

As in previous surveys, a wide majority of farmers (83 percent) report that income *either increased or did not change* after they started farming with fewer purchased inputs. Fewer than 17 percent say net income decreased.

"I paid my farm off two weeks ago, and I sure didn't do this by making the chemical companies rich," said a 350-acre grain farmer in Hicksville, Ohio. "I have banded soybeans for 44 years," he pointed out. He's also done well with up to 50 acres of mammoth red clover the past five years: by harvesting and selling the seed, he grosses about $500 per acre.

# Responded	553
Income Not Changed	249 (45.02%)
Income Increased	206 (37.25%)
Income Decreased	92 (16.63%)

Sources Farmers Use For Information

For information about cutting inputs, farmers rely primarily on farm publications, their own ideas or other farmers. They consult Extension agents, salespeople and other sources, but not to the degree they depend on periodicals or other farmers.

# Responded	754
New Farm Magazine	499 (66.18%)
Farm Publications	385 (51.06%)
Own Ideas	335 (44.42%)
Other Farmers	201 (26.65%)
Books (Usually Old)	146 (19.36%)
Extension Agents	77 (10.21%)
Consultants	65 (8.62%)
Salespeople	48 (6.36%)
Growers' Association	25 (3.31%)
Other	29 (3.84%)

Factors Influencing Net Income

Here, weather was a choice, and was seen as the main factor (by 48 percent of respondents) affecting any change in net income. A close runnerup was use of cost-cutting measures (45 percent), followed by commodity prices (30 percent), yield changes (17 percent) and tillage changes (14 percent). Of these, only weather and commodity prices are generally out of a farmer's direct control.

# Responded	591
Weather	283 (47.88%)
Cost-Cutting Measures	264 (44.67%)
Commodity Prices	177 (29.94%)
Yield Change	99 (16.75%)
Tillage Change	85 (14.38%)
New Crops Added	32 (5.41%)
Weed/Insect Problems	2 (0.33%)
Other	14 (2.36%)

Effect Of Farm Size On Net Income

One might have expected small-scale farmers to have an easier transition to fewer purchased inputs, and large-scale farmers to have more problems, at least as seen by net income changes in the first few years after changing. That wasn't the case, however. Regardless of acreage, the surveys suggest that most farmers' net incomes improve over time after reducing off-farm purchases.

A high-value grower in Michigan said he thought the small size of his operation helped him. "Small farmers have no choice. We have to be smarter to survive," he said. "My net is higher than any of my big neighbors'. My costs are so low that a bad year (I don't have any, really) won't hurt me. I feel sorry for 'traditional' farmers. They are trapped in a self-destructive system."

Chapter 4: Forget Conventional Wisdom

Skeptical farmers recite a litany of worries whenever someone asks why they haven't reduced their chemical use. Yields and income will drop, they say. Pests will take over. Equipment, labor and management needs will increase. And, even worse, commodity bases and government payments will decrease, they fear.

But experienced low-input farmers tell a different — and more optimistic — story. They explain in detail how satisfied they are about regenerating on-farm resources such as soil and water. When you visit their farms and walk their fields, you see their results. We've been visiting these farmers for years, and have featured their success stories in *The New Farm* magazine since 1979.

Mike Reicherts, New Hampton, Iowa, sidedresses strips of corn using a honey wagon he modified to deliver precise rates of liquid hog manure.

In this book, we've profiled the experiences of 18 farm families who reduced their use of off-farm inputs. We've also presented compelling data from surveys of hundreds of *New Farm* FONE ("Farmers Own Network for Education") participants who have successfully cut their chemical use. Most of these farmers say they've produced good yields and that their profits have stayed the same or *increased*.

Many of the sustainable farmers we've talked to say that labor, management and information needs are ongoing concerns. But through experience, diversification and networking with like-minded farmers, most report encouraging short- and long-term results with their new farming systems.

Are their results typical of all U.S. farmers who are reducing their reliance on off-farm inputs? Although some regional differences exist, independent surveys by a number of groups interested in American agriculture support much of what we've found over the years through *New Farm* reader surveys.

Myth vs. Reality

"Conventional wisdom on reduction of inputs is not what it would seem to be," says Kevin Carroll, a former graduate researcher in the University of Wisconsin's (UW) College of Agriculture and Life Sciences. Carroll and Margaret Krome, director of a sustainable agriculture program for the nonprofit Wisconsin Rural Development Center, conducted surveys of two groups of Wisconsin farmers in 1988.

Krome looked at a sample of 389 conventional farmers. Her goal was to determine potential or *perceived* obstacles keeping them from reducing chemicals.

Carroll surveyed 137 farmers who had reduced or eliminated chemicals. "My purpose was to look at what kinds of problems farmers face when they *actually* cut back on chemicals," he says.

Fear of lost yields and/or income is the greatest barrier to low-input farming among the conventional farmers surveyed. "They thought that by reducing chemicals, their profits would drop," says Krome. More than two-thirds of her respondents expressed that concern.

Another fear is that weed and pest problems will be too severe with fewer chemicals. Some 46 percent of Krome's respondents worry about this.

A third perception about cutting chemicals was that it will require a lot of new equipment. More than 45 percent of Krome's respondents think that will be the case, with more than 35 percent listing it among their top three reasons for not reducing their chemical use.

Reality: Profits Increase

"Yield losses weren't a big problem," says Carroll, who surveyed the low-input group. Only 35 percent of his respondents said yields fell when they cut chemicals, and just 32 percent believed yield losses cut some of their profit. "More of the low-chemical farmers (76 percent) showed an *increase* in profits from savings on inputs than loss in profits because of loss in yields." That was true of all farmers in his survey, although cash-grain producers were more likely to report yield losses than were dairy farmers, he notes.

These results could vary in a state with less emphasis on dairying and pasture, Carroll points out. "Wisconsin is tailor-made for sustainable agriculture," he says. Other factors not included in his survey, such as weather, commodity prices and debt load, also could influence results. "In any event, we're not talking about major swings in profitability either way," Carroll notes.

Problems due to weed and insect pressure were much less than what conventional farmers predicted, says Carroll. "There was a marked decline in major pest problems during the input reduction, from 14.2 percent of respondents to 6.3 percent. Some farmers said they were willing to put up with more weeds and insects now that they have reduced pesticide inputs, but these increases in pests are not affecting yields too much."

Getting reliable information about low-input farming seems to be a problem.

Predicted needs for new equipment seem to jibe with reality. More than 45 percent of conventional farmers think they will need new equipment if they cut chemicals, and more than 35 percent say that's a major reason they'd avoid low-input farming. Almost 38 percent of low-input farmers say such purchases are indeed common.

The word "new" could have influenced the results, the researchers believe. Farmers could have taken it literally, to mean "showroom new." Often, low-input farmers simply add one or two items — a rotary hoe, for example — that might be bought used. So exactly what was measured here isn't clear.

Conventional farmers tend to underestimate labor needs attributable to reducing chemical inputs. Almost 44 percent of the conventional group expected labor to increase if they reduced chemicals, with 28 percent of the conventional farmers citing this as one of their main obstacles to cutting chemicals. But 64 percent of the low-input group said labor does increase when you adopt low-input systems.

The increased labor didn't seem attributable to more field operations. Only 16 percent of Carroll's respondents said *more time cultivating weeds* was a major problem when reducing chemical use. Fewer than 8 percent said it was a major problem after they had completed the transition. The percentage is even lower for those saying they spent more time handling manure.

Carroll and Krome think farmers took added management time into account when answering labor questions. "One farmer commented that if you asked him where labor inputs increased, he couldn't point to hours logged on a tractor or spent in a barn," observes Carroll. "What he could say is that he was paying more attention to the condition of his crops and fields — where weeds were looking bad and what he should plant in a particular field the following year."

Loss of base acres wasn't as large a concern as the two expected. Only about one-fourth of the low-input farmers said their crop base acreage declined when they cut chemicals. Even fewer of the conventional farmers (13 percent) said this would be a barrier to reducing chemical use. "But this represents a state [Wisconsin] that has a high percentage of livestock farmers, who traditionally depend less on base acres in determining their crop mix," says Krome. "Farmers in cash-grain states whom I've talked to often say they are concerned about base acres."

Getting reliable information about low-input farming seems to be a problem for both groups. Fewer than 28 percent of the conventional growers

Farmer Surveys Dispel Low-Input Myths

	'90 Reader Survey	'89 FONE Survey	Wisc. '88 (Carroll) low-input actual	Wisc. '88 (Krome) convent'l [perceived]
# Farmer Responses	497	814	137	389
Mean Acres	434	377	148	254
Region				
North-Central	59%	82%	100%	100%
Northeast	13%	18%		
Western	9%	0%		
South	19%	0%		
Main Enterprise				
Cash Grain	33%	64%	9%	100%
Dairy	15%	9%	60%	
Beef	25%	8%	15%	
Hogs	13%	6%		
High-Value	14%	10%	16%	
Total:	100%	97%	100%	100%
Yields:	N/A			
Decreased		30%	35%	[71%]
Increased		7%	18%	
Stayed Same		62%	47%	
Net Income				
Decreased	6%	17%	33%	[68%]
Increased	29%	37%	66%	
Stayed Same	65%	45%		
Labor	N/A			
Increased		51%	64%	[44%]
Decreased		20%		
Stayed Same		28%		
Primary Weed Control		N/A		
Mechanical	47%		62%	18%
Chemical	39%		32%	81%
Crop Rotation	15%			
Cover Crops	8%			
Other Methods	21%		6%	
Total:	130%		100%	99%
Still Use Chem. Fert.	N/A	72%	51%	84%
Still Use Chem. Insect.	26%	N/A	34%	43%
Still Use Herbicides	60%	N/A	68%	86%
Main Changes In Last Few Years				N/A
Reduced Debt	45%	N/A	48%	
Rotate Crops	35%	24%	57%	
Cover Crops and Green Manures	30%	25%	N/A	
Increased Acreage	30%	N/A	N/A	
Reduced Herbicides	27%	53%	98%	
Reduced Chem. Fert.	27%	56%	77%	
Reduced Chem. Insec.	25%	52%	72%	
Chemical Use				
Moderate	62%	39%		100%
Low or Emerg. Use	17%	44%	100%	
Don't Use Chem.	21%	17%		

Sources: — The New Farm *Subscriber Study*, November 1990. (Survey responses from 719 *New Farm* readers.)

— *Farmers' Own Network for Education (FONE) 1989 Special Farmer Survey Results* edited by Anne Schauer, Rodale Institute, 1991. (Survey responses in early 1989 from 843 farmers in Ohio, Illinois, Indiana and Michigan. Funding provided by the Charles Stewart Mott Foundation.)

— *Reduced Chemical Farming in Wisconsin: Problems and Prospects* by Kevin Carroll, University of Wisconsin, 1989. (Masters thesis based on a survey of 137 low-chemical Wisconsin farmers. Yields are for corn grown in rotation. Funding provided by the Wisconsin Department of Agriculture, Trade and Consumer Protection.)

— *Wisconsin Farmers' Constraints to Reducing Agricultural Chemicals: A Survey Analysis* by Margaret Krome, Wisconsin Rural Development Center, 1990. (Based on 389 respondents to a random-sample survey of 1,000 Wisconsin farmers. Funding provided by the Jessie Smith Noyes Foundation and the Wisconsin Department of Agriculture, Trade and Consumer Protection.)

said a lack of reliable advice is keeping them from trying such methods. "And yet throughout the rest of the survey, there are indications that they really don't have the information they need," Krome says.

When asked if they think there are reliable alternatives to insecticides and herbicides, only 30 percent of the conventional group answered yes. The top two sources of reliable information on cutting chemicals, they indicated, were successful low-input farmers and Extension agents. But just 30 percent and 23 percent of the conventional farmers listed those sources, respectively. University experiment stations were considered reliable sources by only 10 percent of the respondents.

Bridging The Info Gap

"It's clear that these farmers are not confident in the sources of information about cutting chemicals," says Krome. Yet two-thirds of the farmers in her survey said it was likely or very likely they would reduce their chemical use if they could count on individual assistance from specialists when making the transition.

An even larger number — 84 percent — said they would cut chemicals if studies showed groundwater in their area was contaminated with farm chemicals. But more than half (54 percent) believed their use of agrichemicals posed no serious or permanent harm to the environment in their community. That's at odds with Wisconsin Department of Agriculture surveys showing that nitrates and herbicide residues are showing up in the water supply statewide.

More than 70 percent of Carroll's low-input group had problems finding reliable information during their transition to reduced chemicals. "The reduced-chemical farmers are relying on sources outside the conventional information network. These sources are not always satisfactory. But over time, their information problems decreased," says Carroll.

Krome says her survey results suggest that conventional farmers "need practical, locally relevant information on methods they can use to reduce their chemicals." Farmers also need additional information on actual environmental, health and other consequences of their current chemical-use practices.

Fortunately, "farmers have a wealth of knowledge among themselves," says Krome. "But they need to communicate, and push their public institutions to facilitate the exchange of that information and provide accurate information on specific techniques."

Pastures Beat Confinement

Grass-based dairies are more profitable than confinement dairy feeding systems and require *less* labor, researchers at the University of Wisconsin (UW) and the Wisconsin Rural Development Center have shown recently. They base their findings on crop, dairy and whole-farm data gathered in Wisconsin over a three-year period at 24 dairies — 13 conventional farms where land type, herd size and production are similar to 11 other dairies that rely on intensive, rotational grazing.

The USDA/SARE-funded study demonstrates that more is not necessarily better. The pasture-based dairies are smaller than the conventional ones, with lower overall milk production per cow.

To adequately reflect calf and cull sales in the whole-farm economic analysis, the researchers used a "milk equivalent per cow" figure when calculating per-hundredweight returns and cost-efficiency measures. They calculated the milk equivalent per cow by dividing total milk, calf and cull sales by the average price of the main product. On that basis,

Pasture-based dairying provides Charlie Opitz, Mineral Point, Wis., with solid profits, even with 800 milkers.

Wisconsin Dairy Study 1991	Pasture-Based	Conventional Farms
# Dairies Studied	11	13
Mean Tillable Acres	185	235
Mean Number of Cows	47	52
Production Per Cow (Cwt)	158	202
Milk Equiv. Per Cow (Cwt)	181	232
Total Labor Hours	4,039	4,999
Fertilizer Costs/A	$1.89	$21.11
Herbicide and Insecticide Costs/A	$0.61	$ 7.83
Total Pest. & Fert./A	$2.50	$28.94
COST/COW/YEAR:		
Pesticide/Fertilizer:	$ 10	$ 133
Fuel/Repairs:	138	228
Crop Production:	79	124
Purchased Feed:	446	428
TOTAL FEED COST		
Per Cow	$ 673	$ 913
Per Cwt Equiv.	3.72	3.93
Vet/Breeding/Misc.	324	426
Labor Cost	431	484
Other Fixed Costs	659	886
TOTAL COST		
Per Cow	$2,087	$2,709
Per Cwt Equiv.	11.52	11.68
INCOME/CWT EQUIV./YEAR:		
Average Return	$12.14	$11.83
Net Return	0.62	0.15
Net Cash Return	6.64	6.06
NET RETURN/COW:	$ 111	$ 36
Dairy Enterprise*		
Gross Income/Farm	$103,259	$142,740
Net Income/Farm	$ 5,217	$ 1,820
Net/Gross Ratio	5.1%	1.3%

Source: "Whole-Farm Economic Analysis of Medium-Sized, Single-Family Dairy Farms That Differ in Their Use of Purchased Chemical Inputs," by Marvin Kamp *et al*, Wisconsin Rural Development Center Inc., 1991.
***Note: Low milk prices contributed to a significant loss of income for all dairy farmers in 1991.**

pasture-based milk production was 47 cents per hundredweight more profitable than conventional production.

The dairy operators relying on rotational grazing report greater flexibility in the use of on-farm time, a reduction in labor needed during the grazing season, and an improvement in their quality of life. Conventional farms incurred 960 hours of additional labor per year, compared with pasture-based farms. That's about 18 hours more per week!

Pasture dairies spent an average of $26.44 less per acre on synthetic pesticides and fertilizers. Eliminating continuous corn in the rotation and plowing down alfalfa stands more regularly made it easy for these dairies to reduce their off-farm inputs. "Pasture-based farms carry this cost-cutting further by shifting a portion of the rotation to pastures," notes Dr. Richard Klemme, professor of agricultural economics at UW and co-author of the economic analysis.

Purchased feed costs were slightly higher for the pasture-based dairies than for conventional dairies.

That could change over time, as pasture-based dairy operators acquire additional experience with rotational grazing and further reduce their off-farm grain purchases. *Total* feed costs, however, clearly favor the pasture dairies ($3.72 per hundredweight equivalent, compared with $3.93 for conventional dairies). "Even with lower milk production per cow, this suggests lower production costs for the overall enterprise for the pasture-based farms," Klemme points out. Although pasture-based dairies were more likely to use mechanical cultivation instead of herbicides for field crops, they also showed lower investment in crop-production equipment.

The more sustainable the farmer, the more gross income he or she is likely to retain.

With more cows and higher production, the conventional farms show higher gross incomes. But because of significantly lower input costs, the pasture-based dairies in this study earn considerably more per cow — $75 additional net per head, or $0.47 *more* per hundredweight. Clearly, it pays to put your cows on pasture, rather than feeding high-input grains.

Multistate Evidence

An ambitious research program sponsored by the Northwest Area Foundation (NWAF), land-grant universities and sustainable ag groups in the North-Central and Northwestern U.S. points out a number of important features about sustainable farming in America as of 1990:

❖ Sustainable farmers generally operate fewer acres than conventional farmers.

❖ Sustainable farms are more diversified and have much more acreage devoted to crops other than government-subsidized commodities than do other farms.

❖ Sustainable farms typically produce comparable or *higher* net incomes on a per-acre basis than do conventional farms, although conventional farms have higher gross incomes.

❖ Transitional farmers experience more economic stress than do experienced sustainable farmers or conventional farmers.

❖ Sustainable operations generally need more labor per acre than do conventional operations.

❖ Sustainable farms tend to be family operations that rely less on hired labor than do conventional farms.

Let's consider a few of the findings from the Northwest Area Foundation Initiative. As part of this study, researchers from the University of Minnesota and Land Stewardship Project studied survey responses from two different groups: 453 randomly selected farmers in the state and 374 Minnesota farmers who belong to sustainable agriculture groups. They ranked farmers on a scale of sustainability, categorizing them as either high, medium or low, depending on their actual farming practices and attitudes about chemical use and crop rotations.

Farmers who were higher on the sustainability scale tend to run smaller operations (361 acres on average), while lower-ranked farmers have more land (623 acres on average). In most cases, the difference is in acres *rented*, rather than in acres actually owned.

The study shows that "low-sustainable" farmers in Minnesota plant twice as many acres of corn as do "high-sustainable" farmers, and almost four times as many acres of soybeans. The more sustainable farmers tend to have more acres of small grains, alfalfa, pasture, woodlands and wetlands.

More of the highly sustainable farmers (88 percent) raise livestock than do the "low" sustainable group (42 percent). Statewide, about 65 percent of Minnesota farmers raise some livestock. Of the farmers without livestock in the survey, more than half were in the low-sustainability group.

"The most conventional farms are more likely to have large market-hog operations, averaging 356 hogs, compared to an average of 21 on the most sustainable farms," the researchers found. Sustainable operations were more likely to have dairy cows, cattle or other livestock, and in smaller herds than do conventional farms. Fifty-one percent of "high-sustainable" farmers use rotational grazing, compared with just 26 percent of the statewide survey.

The more sustainable the farmer, the more gross income he or she is likely to retain, the Minnesota study shows. "The least-sustainable operations had higher median gross incomes ($105,999) than the highest sustainable category ($58,750), but more

Sustainable Agriculture in Minnesota, 1990

	Statewide	Low Sustainable	Medium Sustainable	High Sustainable
# Surveyed	453	125	155	107
Mean Acres	459	623	400	361
Main Enterprises (Acres/Farm)				
Cropland	378	557	319	262
Pasture	40	17	40	66
Other	41	49	41	33
Have Livestock:	65%	42%	71%	88%
Mean # Livestock				
Beef	42	30	47	42
Dairy	19	4	19	29
Swine	192	395	224	43
Poultry	7	0	12	4
Other	14	5	16	17
Main Crops (A/Farm)				
Corn	132	189	115	94
Soybeans	97	168	80	43
Oats	21	11	23	28
Alfalfa	33	10	37	53
Other	58	91	47	38
Median Net Income (Per Farm)	$17,510	$22,836	$13,562	$17,451
Net/Gross Income Ratio	22%	22%	18%	30%
Median Govt. Payments	$6,258	$9,214	$5,857	$3,597
Govt. Payments as % Net Farm Income	36%	40%	43%	21%
Labor:				
Family Employees	1.5	1.5	1.3	1.7
Non-Family Emp.	1.7	1.7	1.5	0.9
Future Expectations				
Nutrient Use				
All-Chemical	26%	62%	15%	2%
50:50	49%	33%	64%	45%
All-Natural	24%	5%	21%	53%
Weed Control				
All-Chemical	15%	33%	10%	3%
50:50	61%	58%	70%	49%
All-Natural	24%	9%	20%	48%

Source: *Sustainable Agriculture In Minnesota: Report Phase 1* by Dr. Dario Menanteau-Horta *et al,* Center For Rural Social Development, University of Minnesota, 1991. (Based on random-sample survey responses in 1990 from 453 Minnesota farmers. Study funding provided by the Northwest Area Foundation.)

comparable net incomes ($22,836 and $17,451)," reports Cheryl Miller, who wrote the Northwest Area Foundation report. With a 30-percent ratio of net-to-gross farm income, high-sustainable farmers are generating products with a better financial margin than are low-sustainable farmers, who retain only 22 percent of gross income, on average.

Not surprisingly, high-sustainable Minnesota farmers rely less on government subsidies (21 percent of their net income in this survey) than do low-sustainable farmers (40 percent of net income). Those figures include CRP, deficiency payments and drought relief, but not commodity loans. Most of the sustainable farmers, of course, have fewer acres in cropland than the conventional farmers have, and many government payments are tied to crops.

Only 40 percent of the Minnesota farmers in the survey report using reduced-tillage practices, with 52 percent saying they have tried and abandoned some form of no till, minimum till, ridge till or a related practice. Researchers found no significant difference in tillage practices among the three levels of sustainable farmers. Tradition and equipment expense could have been reasons for this, the researchers believe.

Nitrogen crediting and budgeting, however, was related to degree of sustainability, with higher sustainable farmers more likely to take legumes, manure, crop rotation and soil testing into account when determining their N purchases. Fifty-five percent of the low-sustainable farmers report never using (or using but abandoning) methods of accounting for their nitrogen use.

Nearly half of the farmers in the statewide survey use equal amounts of synthetic and natural fertilizer, with 75 percent using some form of purchased N. But 59 percent indicate they are trying to reduce their reliance on chemical fertilizers and pest control. Twenty-six percent think they will be using *more* synthetic N in the next five years; 24 percent expect to increase their use of natural fertilizer. While 54 percent of the "high-sustainable" farmers hope to be using only natural forms of fertilizer in the next five years, only 5 percent of the "low-sustainable" farmers expect that to be the case.

High-sustainable farmers are more likely to use cultivation and rotation to control weeds. Almost half of them expect to eliminate herbicide use, while 91 percent of the low-sustainable group expect to continue relying on some herbicides.

A majority (59 percent) of the farmers report having changed some of their farm practices due to a labor problem. Low-sustainable farmers tend to use more chemicals, buy more machinery, switch to minimum tillage or hire migrant help. Mid-ranking farmers tend to reduce their number of livestock, buy machinery and reduce tillage. Many of the high-sustainable farmers simply reduce their acreage and rely more on family labor.

From The Iowa Heartland

Sustainable farms in Iowa also tend to be smaller than conventional farms — 391 acres vs. 673 acres — according to 1989 survey data collected for the Northwest Area Foundation by researchers at Iowa State University and by Practical Farmers of Iowa. Labor figures weren't available when we went to press. But other data mirrored trends elsewhere. Compared with conventional Iowa farmers, sustainable ones rely more on:

❖ livestock (91 percent vs. 58 percent)

❖ small herd sizes

❖ forage and pasture (22 percent of acreage vs. 4 percent)

❖ small grains (13 percent of acreage vs. 1 percent)

❖ non-chemical weed control and fertilizer

"Sizable differences were found in nitrogen fertilizer use among the farm types," NWAF writer Cheryl Miller observes. "Only 1 percent of the conventional farmers, compared with 91 percent of sustainable farmers, applied less than 100 pounds of fertilizer per acre in selected cornfields in 1989." That year, she notes, "Conventional farmers set the highest yield goal for corn (150 bushels per acre). Transitional goals were somewhat lower (140 bushels per acre), and sustainable farmers set goals lower still (126 bushels per acre). Actual yields were considerably more alike than goals, with conventional farms at 131 bushels per acre, transitional farms at 122 bushels per acre, and sustainable farms at 124 bushels per acre."

Self-ratings, in which farmers rated their former, current and future nutrient use and weed-control practices, show a clear trend toward using fewer chemicals, even among conventional farmers. Transitional farmers report the greatest shift away from chemical use, with sustainable farmers — many of whom already had reduced their chemical use — not far behind. Conventional farmers report less change, but they too indicate they expect to continue cutting some of their chemical use in the next five years.

Gross income generally was higher for conventional farms than for sustainable ones, but that was largely a function of farm size. Sixty-one percent of conventional farmers report gross income in excess of $100,000, vs. 44 percent of sustainable ones. Debt-to-asset ratios were comparable, except for transitional farmers, who report a higher debt load. That could be due to the purchase of additional tillage equipment. (Or, it could be the reason they're making the transition to sustainable methods in the first place!) As part of the study's next phase, Iowa researchers will be looking more closely at farm income.

Survey of Iowa Farmers, 1989

	Conventional	Transitional	Sustainable
# Surveyed	120	74	116
Mean Farm Size (Acres)	673	578	391
Main Enterprises (Acres/Farm)			
Cropland	639	514	289
Forage/Pasture	27	58	86
Other	7	6	16
Main Crops (Acres/Farm)			
Corn	410	312	152
Soybeans	222	173	86
Small Grains	7	29	51
Forage/Pasture	27	58	86
Other	7	6	16
Have Livestock	58%	85%	91%
% Having Livestock/Mean #			
Feeder Cattle	27% / 199	39% / 182	45% / 93
Cow-Calf	15% / 39	36% / 49	40% / 46
Dairy/Herd Size	1% / 50	8% / 59	22% / 58
Farrow-To-Finish/# Sold	28% / 1438	39% / 1067	35% / 805
Feeder-To-Finish/# Sold	12% / 1420	28% / 984	13% / 583
Feeder Pigs/# Sold	6% / 888	11% / 825	14% / 710
Sheep/# Sold	3% / 55	13% / 48	24% / 41
Gross Income Per Farm			
Under $60,000	18%	21%	32%
$60,000 To $99,999	21%	17%	24%
$100,000 To $199,999	34%	37%	29%
$200,000 To $299,000	11%	13%	8%
$300,000 Or Higher	16%	12%	7%
Debt/Asset Ratio	32%	47%	31%
Nutrient Use (1 to 10 scale, where 1 = all-synthetic; 10 = all-natural)			
1984 Self Ratings	3.0	3.0	5.6
1989 Self Ratings	3.1	4.6	6.8
1994 Expected Ratings	3.9	6.5	7.6
Weed Control (1 to 10 scale, 1 = all-herbicide; 10 = tillage and other cultural control)			
1984 Self Ratings	3.6	2.9	5.6
1989 Self Ratings	3.7	5.1	7.0
1994 Expected Ratings	4.3	6.7	7.7

Source: *Transition to Sustainable Farming Practices in Iowa* by Dr. Gordon Bultena *et al*, Department of Sociology, Iowa State University, 1993. Sampling based on farms having a minimum of 50 acres in 1989 — that excluded 18 percent of Iowa farms — and at least some corn acreage. Funding provided by the Northwest Area Foundation, ISU's Iowa Agricultural and Home Economics Experiment Station, the Leopold Center for Sustainable Agriculture, and Practical Farmers of Iowa.

In North Dakota, where a typical farm exceeds 1,000 acres, "gross farm income and net farm income measured on a per-acre basis were not statistically different," says Dr. David Watt, a researcher at North Dakota State University. "When examined on a regional basis, the conventional farms generally had higher average values for assets, operating ratios, gross and net farm income, on a per-farm and per-acre basis. But sustainable farmers in the eastern region had significantly higher net incomes per farm and per acre."

Wheat farming and cattle ranching increase as

one heads west onto the North Dakota prairies. On a per-acre basis, family labor was 50 percent higher on sustainable farms than for conventional farms, which incurred significantly more non-farm labor expenses ($3.55 per acre vs. just $0.70 per acre). But overall labor figures for conventional farms in North Dakota were smaller than for sustainable or mixed-type farms.

Some of the first-phase conclusions of the Northwest Area Foundation Initiative are worth noting:

❖ "The tendency of sustainable farms to be smaller in acreage, less expensive to operate, and earn comparable net incomes per acre suggests that sustainable agriculture may be an economically viable way to attract and keep more farmers on the land."

❖ Sustainable farmers "report a generally positive economic picture despite an inadequate infrastructure of markets and services and a general lack of government support in the form of commodity subsidies, agricultural research, tax incentives, etc."

❖ "A more varied food supply would result from broad conversion to sustainable practices. Harvest yields of any single crop (such as corn or wheat) would decline, not because yields per acre would drop but because more legumes, alfalfa and other soil-building crops would be added to rotations."

❖ "Better training through agricultural Extension programs may be needed to enable farmers to address ... increased management demands per acre on sustainable farms."

❖ "The multistate surveys point up the economic hardships faced by transitional farmers and underscore the economic advantages provided by low-input sustainable systems."

Regional Trends

Wondering how you compare with *New Farm* readers in your region? Or what they've been doing to prepare themselves for farming in the coming years?

Although close to 30 percent of our readers as a whole say they have increased their acreage in the past five years, the average farm size dipped slightly since our last reader survey in '87. North-Central farms went from an average of 441 to 427 acres, Western farms from 446 to 411 acres and Southern farms from 311 to 277 acres.

The exception is in the Northeast, where mean farm size increased from 176 to 182 acres. Thirty-seven percent of our Northeastern readers say they have increased their farm size in the past five years. That surprised us, given the rate of suburban encroachment in that area.

Livestock and pasture can be found on the majority of farms operated by *New Farm* readers in all four regions. Beef cattle is the most popular livestock enterprise with our readers, followed by hogs and dairy cattle.

About two-thirds of our readers produce hay, although that's true for less than half of the Western

Better Water in Big Spring, Iowa

Since 1981, farmers in a 103-square-mile area of northeastern Iowa have cut their nitrogen use dramatically — without yield losses. Spurred by public health concerns due to high nitrate levels in the groundwater of the Big Spring Basin drainage area, more than half of the 200 farmers there voluntarily reduced nitrogen fertilizer applications. Between 1981 and 1991, average N use for all corn grown in the area dropped 31 percent, from 174 pounds per acre to 120 pounds, saving each reduced-input farmer about $2,400 annually.

Demonstration plots sponsored by Iowa State University (ISU) and a consortium of other groups helped convince many of the farmers to lower their fertilizer rates. Study results encouraged farmers to credit nitrogen contributions from manure and legumes. "Every year since 1985 when the plot work began, producers have lost money by applying purchased N over 50 pounds per acre on corn following alfalfa," says Roger Koster, ISU Extension project coordinator. "Most years, either starter fertilizer or zero N was the most profitable."

In '81, area farmers applied an average of 123 pounds of N per acre on first-year corn following alfalfa. In '91, the rate averaged just 59 pounds per acre — a 52-percent reduction, says Koster.

Other plots showed nutrient needs can be met with manure. "In a six-year study, continuous corn yields in the last three years ('89 to '91) from low rates of surface-applied hog manure (90 pounds of N per acre) averaged 156 bushels per acre, compared with 154 bushels per acre for corn receiving either spring or fall applications of 180 pounds of N per acre from urea," Koster notes. Seventy-six percent of Big Spring Basin farmers credited N contributions from on-farm manure use in '91, he adds.

Nitrate levels in the region are well below health-advisory levels now. Another plus for groundwater quality: Almost 40 percent of Big Spring farmers have reduced their P and K fertilizer use, while 23 percent have cut their herbicide use.

Statewide, Iowa farmers are saving more than $40 million annually by reducing nitrogen fertilizer use for corn. The fertilizer rate reductions amount to more than 200 million pounds annually, according to the Iowa Department of Natural Resources.

The New Farm Subscriber Study, November 1990

	Northeast	North-Central	South	West
Acres Farmed	182	427	277	411
Farm Sales	$66,000	$79,000	$39,000	$96,000
Gross Per Acre	$363	$185	$141	$236
Household Income	$52,000	$51,000	$54,000	$56,000
Main Products	vegetables, dairy, hay	cash-grain, beef, hogs	beef, cash-grain, veg.	vegetables, beef, fruit
Currently Raise Livestock:	58%	67%	71%	53%
(% / # per year)				
Beef Cattle	27% / 29	44% / 57	51% / 45	31% / 67
Hogs	12% / 51	25% / 162	11% / 113	7% / 28
Dairy	32% / 43	13% / 61	6% / 71	9% / 46
Poultry	18% / 89	14% / 95	15% / 60	16% / 60
Sheep	10% / 69	13% / 63	9% / 81	17% / 38
Other	8%	7%	12%	7%
Crops: (% / acreage)				
Small Grains	44% / 41	69% / 67	40% / 69	33% / 116
Corn	55% / 61	71% / 120	44% / 49	13% / 130
Soybeans	14% / 69	51% / 110	18% / 80	0% / 0
Hay	67% / 80	72% / 68	63% / 79	43% / 128
Pasture	69% / 53	70% / 109	83% / 87	55% / 272
Vegetables	61% / 3	39% / 2	61% / 3	57% / 4
Small Fruits	44% / 1	22% / 2	42% / 2	48% / 4
Tree Fruits	40% / 21	29% / 19	47% / 24	44% / 34
Wood	66%	39%	50%	21%
Diversified in the past 5 years?	20%	35%	37%	33%
Most-used legumes and grasses (%)				
Alfalfa	51	65	15	37
Clover	33	24	34	12
Timothy	52	18	16	16
Orchardgrass	32	18	24	22
Fescue	7	17	42	12
Brome	18	23	2	13
Ryegrass	21	9	30	8
Buckwheat	22	8	7	2
Chemical Use (%)				
Intensive	0	4	3	6
Moderate	44	62	38	28
Emergency	21	17	25	28
Never	35	18	34	38
If reduced chemicals, effect on net income				
Increased	21%	30%	24%	0%
No change	73%	64%	60%	94%
Decreased	6%	6%	16%	6%
Main changes in past 5 years (%)				
More cover crops	28	21	29	14
More green manures	37	29	24	30
More acreage	37	24	28	19
Reduced debt	24	42	39	29
Less chem. fert.	22	27	21	11
Less herbicide	26	30	27	14
Less insecticide	23	24	20	16
Better crop rotation	30	35	25	29
Increased mech. weed control in past year	22%	31%	31%	28%

Source: Mail survey of 995 *New Farm* readers in Nov. '90. Rodale Press Market Research Dept. prepared the survey; Data Probe Inc., New York, N.Y., tabulated and reported the 719 responses.

farmers in the survey. The most popular legumes and grasses grown, for whatever reason, by our readers include: alfalfa (59 percent overall), clover (25 percent), timothy (22 percent), orchardgrass (21 percent), fescue (19 percent), brome (17 percent) and ryegrass (14 percent).

Overall, 37 percent say they have diversified their crop or livestock mix in the last five years. More than a third of Southern and Western readers say they have increased their vegetable acreage. The North-Central region is the only area where half or more of our readers don't grow any vegetables.

Of those readers who say they have cut back on any crops, 39 percent have reduced or eliminated corn acreage. Twenty-five percent cut back on soybeans.

More of our North-Central and Western readers report having cut chemicals in the past five years, compared with Northeastern and Southern readers. Overall, 38 percent of our readers say they either do not use farm chemicals or use them only in an emergency. But that's true for 66 percent of our Western readers, vs. just 35 percent in the North-Central region. That could mean more farmers in the Western U.S. have already reduced their chemical use, perhaps due to stricter environmental regulations in some states such as California.

We see a trend toward increased reliance on mechanical weed control in most regions.

Every survey that we've done shows that farmers cut chemicals for three equally important reasons: environmental, financial and health. Very few do so for organic premiums.

A whopping 94 percent of our readers, however, report their income has either stayed the same or *increased* after they cut chemicals. Our Southern readers haven't fared quite as well — 16 percent say their income has declined since reducing their use of agrichemicals. Income per acre in the South is also the smallest of the four regions, followed by the North-Central region. In the Northeast, where farms are smaller, income per acre is higher. Of course, we'd expect our readers as a whole to be better able to do a good job of cutting chemicals than, say, a national sample of farmers where many haven't been reading *The New Farm.*

We see a trend toward increased reliance on mechanical weed control in most regions. Twenty-two percent of our Northeastern readers say they rely more on rotary hoes and cultivators than they did the previous season (vs. 18 percent in our '87 survey). Comparable figures for other regions are: 31 percent of North-Central farmers (vs. 26 percent in '87); 31 percent of Southern farmers (vs. 23 percent); and 28 percent of Western farmers (a slight drop from the 29 percent in '87).

Debt reduction was a priority for many *New Farm* readers in all regions during the past five years, but especially in the North-Central region and the South. Better crop rotations and increased use of green manures and cover crops were reported by farmers in all regions.

Most of our readers know that one of the best ways to maintain profitability over time is to improve the health of their soil. When Bob Rodale established *The New Farm* magazine in 1979, he wanted it to provide farmers with practical information that would do more than just maintain the productivity of their farms. "Sustainable agriculture means to expand the goal of agriculture beyond just producing food and fiber at a reasonable cost," he said. "We have to do all those things and more — to also farm *regeneratively*, making the land and the rural environment better with each passing season."

Chapter 5: Low-Input Systems Really Work

Organics To The Test

KUTZTOWN, Pa. — You can reduce your use of off-farm inputs in a shorter period of time than you might think. This transition is possible even where chemical fertilizers and herbicides have been used for years to maintain yields at the expense of soil health. You can succeed even if you lack ready access to livestock manure. What's more, you can lower your herbicide use even if you're reducing tillage.

In replicated studies since the early '70s, Rodale Institute researchers have shown how the use of crop rotations, legumes and other cover crops, and chemical-free farming practices can provide all the nutrients and pest control needed to harvest profitable yields. Through field days, "Take Charge" workshops, *New Farm* articles and other teaching methods, Institute staff have been helping farmers regenerate their soil while cutting chemicals.

Can you do this overnight? No, you can't. Research shows that transition time depends partly on your farming system and crop choice, soil and land characteristics, climate and ease of obtaining — or growing your own — natural fertilizer.

As with quitting a habit such as smoking, transition from chemical-dependent farming can cause some discomfort during the period when you're improving your soil health and gaining experience with alternative methods. But just as quitting smoking saves money and lengthens lifespans, so too will reduced-chemical farming methods save upfront expenses, sustain long-term profits and improve your farm's most important assets: soil, water and other natural resources.

To help your transition — or get you thinking about further refinements if you've *already* pared your chemical use — let's examine some findings of two ongoing studies at the Rodale Institute Research Center (RIRC) near Kutztown, Pa. The first is called the Farming Systems Trial (FST). The other is the Low-Input Reduced Tillage study (LIRT).

Initially called the "Conversion Project," FST began in 1981 as a feasibility study of chemical-free farming. It compares a conventional cropping system (one using a corn-corn-soybean rotation and following Extension recommendations for synthetic fertilizers and pesticides) with transition to either of two low-input systems: one with livestock and one without.

By "low-input," the researchers meant going without *purchased* off-farm inputs such as synthetic fertilizer and pesticides — not going without *any* inputs. These low-input, organic systems emphasize increased use of on-farm resources including operator management.

The "Low-Input with Animals" (LIA) system

RIRC Photo

Established by Bob and Ardie Rodale in 1971 and now internationally known, the 333-acre Rodale Institute Research Center near Kutztown, Pa., sponsors many regenerative farming studies.

Farming Systems Trial Rotations 1986 – 1990

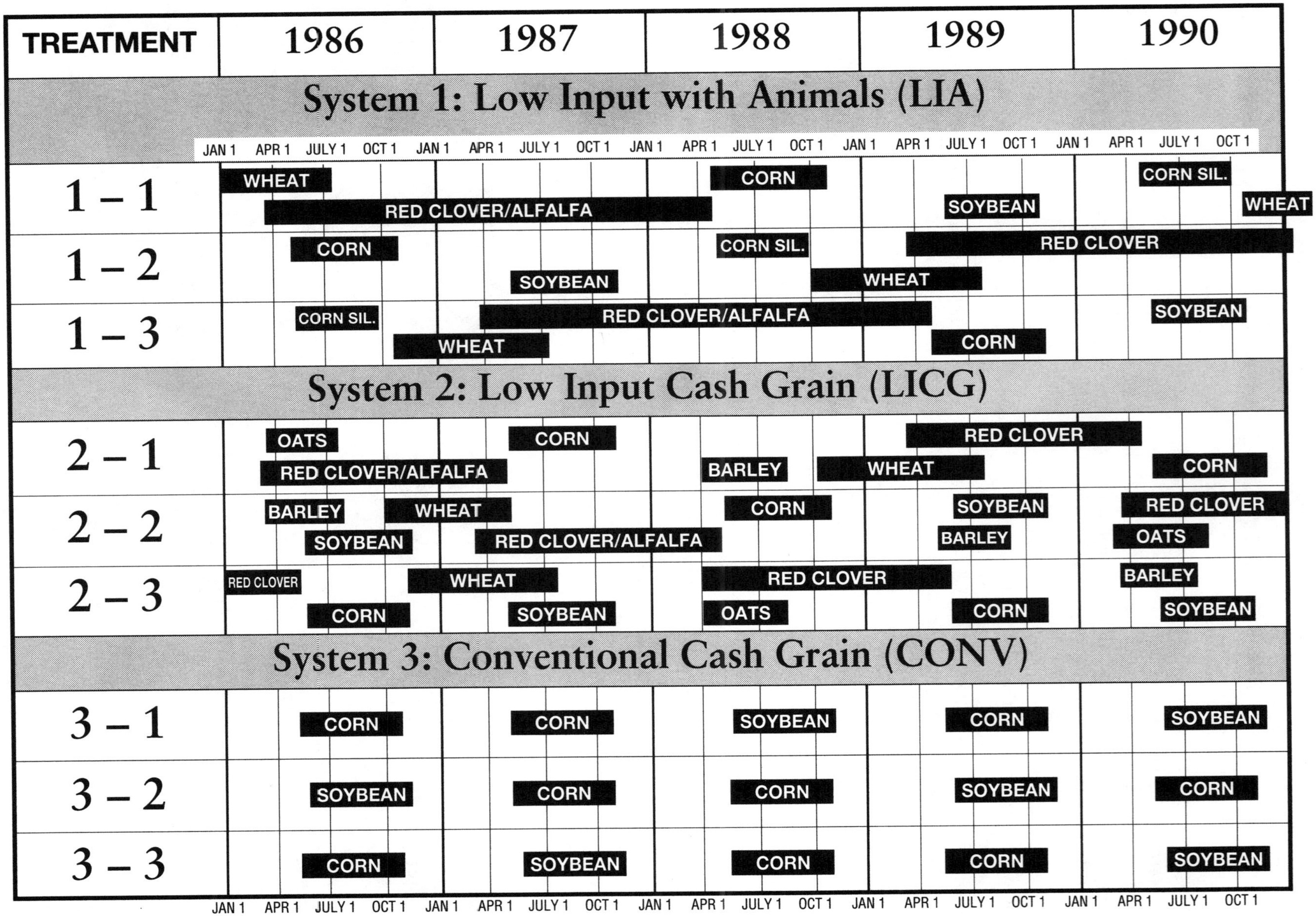

includes hay and silage crops in rotation with grains and use of cattle manure as a nitrogen source for corn. The diversified five-year rotation includes corn grain, soybeans, corn silage, winter wheat and red clover/grass hay.

The Low-Input Cash-Grain (LICG) system features legume green-manure crops as the primary source of nitrogen for three-year rotations of corn, soybeans and small grains. It includes a cash-grain crop each year. Instead of herbicides, the low-input systems rely on mechanical cultivation, cover crops and crop rotation for weed control.

RIRC scientists started each rotation at different entry points to see how that would affect the transition and economics of each system. "We intentionally started with corn in some rotations — something we didn't think would work," says Dr. Rhonda Janke, RIRC director. "That way, farmers could learn from our best treatments and also from our worst treatments. We wanted farmers to learn from our mistakes, not make them themselves."

By renting a neighbor's 15-acre, shaley silt-loam field that had been farmed with chemical-intensive methods for years, RIRC staff made sure that the randomly assigned treatment plots provided a realistic test of transition. During the study's first four years, researchers measured a noticeable "transition effect" in low-input plots, in which below-optimal soil conditions adversely affected plant growth.

"From 1981 to '85, we had less net return per acre in the organic rotations, because we *did* go cold-turkey on this transition," says Janke. "It took approximately four years to get good, consistent yields. It was partly biological — we were building up the N pool — and partly management, making the switch from herbicides to cultivation and other cultural practices.

RIRC Photo by Anthony Rodale

Seeded with orchardgrass into wheat, red clover supplements cattle manure as an N source in LIA plots.

"In the LICG system, the rotation that started off with small grain interseeded with clover did the best. The least profitable was the one that started with corn. If you are growing corn, it's not going to work without added N."

Starting with corn produced average net earnings of just $112 per acre per year in the LICG system from 1981 to 1985. Even the worst conventional rotation netted $26 *more* per acre annually during this period.

Increased weed pressure following a switch from herbicides to mechanical weed control also contributed to initial yield declines in organically grown corn, which averaged about 80 bushels per acre during the transition. County average yield for corn in southeastern Pennsylvania ranges from 90 to 110 bushels per acre.

A Quick Turnaround

By 1984, the low-input plots showed dramatic improvement. In FST's fourth year, Institute Chairman Bob Rodale already had reason to call the study "our pride and joy." "Results are impressive," he noted. "Crop yields are fine in the organic plots. Tillage is noticeably easier compared with turning the soil in the chemical plots.

"The conversion experiment shows that organic farming can be the most profitable kind of agriculture. Business-minded farmers are taking note." So was the USDA, which had sent a soil scientist to join the FST research team, he added.

"Since 1984, corn yields in the low-input systems have consistently matched or exceeded conventional yields of 140 bushels per acre," reports Steve Peters, project leader for the study, which was renamed the Farming Systems Trial in '86. In 1985, corn in LICG yielded 152 bushels per acre, compared with

Farming Systems Trial Rotations 1991 - 1995

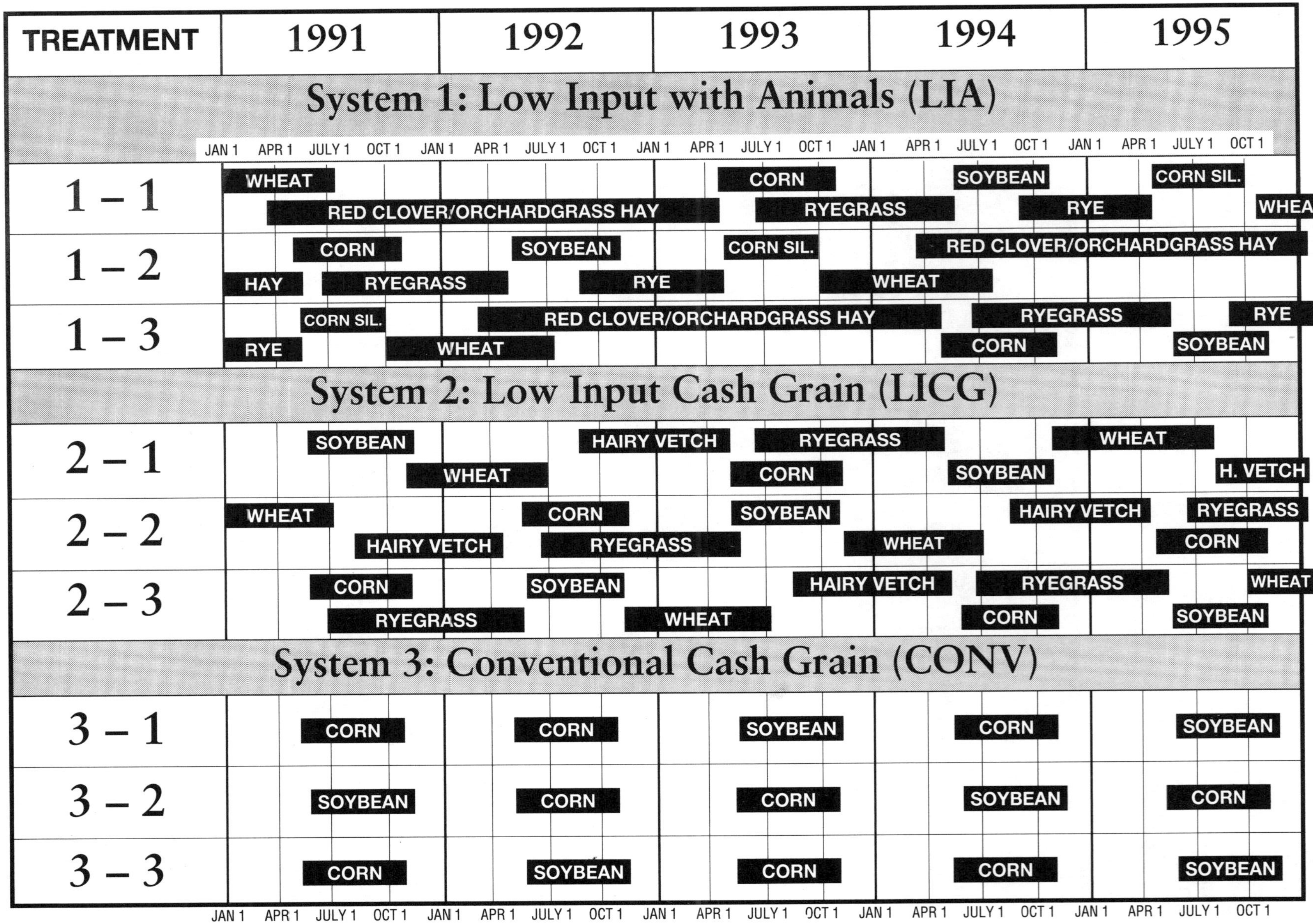

RIRC Photo

Soybeans in FST's low-input plots were almost weed-free even before second cultivation in July '90.

136 bushels for conventional corn.

"For soybeans, we've matched or exceeded county average (anywhere from 30 to 50 bushels per acre) in all systems. That's despite higher weed pressure in the low-input plots than in the conventional ones," he adds.

All told for the transition period from '81 to '85, returns to land and management per year averaged $185 per acre in LIA, $156 per acre in the conventional cash-grain system and $133 per acre in LICG.

Conversion Tips

Based on their transition findings, RIRC researchers make these recommendations to farmers who want to cut their chemical use:

❖ Begin with legume hays or soybeans, or with fields having good fertility.

❖ Avoid starting with corn.

❖ Allow a three- to five-year period for yields to stabilize for N-demanding crops such as corn.

❖ Eliminate monocultures.

❖ Include legume green manures in the new rotation, usually before crops that are high N users, such as corn, cotton or sorghum.

❖ Alternate cool-season crops such as hay and small grains with warm-season crops such as corn and soybeans, to interrupt weed cycles.

❖ When eliminating synthetic fertilizer and pesticides, don't go cold-turkey on a large scale, since that could be financially risky during transition.

❖ Experiment on a small scale so you can gain experience with practices such as rotary hoeing or overseeding, and reduce risk.

❖ Be flexible and adapt to the changing conditions on your farm.

You'll notice the recommendations don't say to invest in livestock if you lack a ready supply of manure. Although many farmers boost their net income and reduce their crop risks by diversifying with a livestock operation, you needn't take on the added labor of managing a herd or a flock just to help get you through transition.

"Low-input farming without livestock is possible and profitable," says Janke. "The transition doesn't take four years if you already have a diverse rotation. And it certainly doesn't take four years to start *cutting back* on purchased inputs. The key is to learn about your system before you jump into it. Start with one field at a time and develop the crop rotation while cutting back on inputs slowly."

The RIRC staff use legumes as the sole source of N in the LICG rotation. You might find it easier to sidedress N in corn based on a late-spring soil nitrate test. As you increase a field's fertility with an improved rotation, you can gradually cut back on the sidedress, if soil-test results allow.

Be sure to set reasonable yield goals too, suggests

Take Charge Of Income

Financial risk diminishes the longer you farm with fewer purchased inputs. And you won't watch profits see-saw year to year as much as you would if you were farming conventionally.

Those are two of the conclusions of a whole-farm financial study[1] coordinated by Dr. Jim Hanson, a farm management specialist at the University of Maryland. Hanson based his study on data from the Rodale Institute's Farming Systems Trial (FST).

The financial analysis compared the FST's Low-Input Cash-Grain system and its conventional cash-grain system. The researchers analyzed crop yields, prices and costs during the period 1981 to 1989. They plugged these figures into scenarios representing a low-input, 750-acre grain farm in the Mid-Atlantic region and a conventional grain farm with the same amount of tillable land. They assumed custom operators did all harvesting and fertilizing, and seasonal hired labor supplemented 250 hours of operator and family labor per month.

Hanson and his team factored in other expenses that farmers in similar situations would have incurred, such as income taxes and debt financing. They also considered the effect that government farm-support programs would have had on each system.

Among their findings:

❖ Variation in profits is far greater for conventional farms than for low-input ones. Without federal commodity programs, the low-input scenario showed profits would exceed $16,166 in three out of four years. *With* government support, a comparable conventional system would show profits only in excess of $12,777 in three out of four years. Much of the credit for less variability in the low-input system came from its diverse rotation — on average, one-third of the gross income came from crops other than corn and soybeans.

❖ Cash operating expenses for the low-input system were 26 percent lower than for the conventional system. That mostly reflects the costs conventional farmers incur for synthetic fertilizers and pesticides.

❖ For *total* profits during this period, the conventional system with government programs comes out ahead. This scenario exceeded a low-input one (that assumed some government support of corn acreage) by almost $7,000 per year ($39,193 vs. $32,464).

❖ Profits trend upward for the low-input system. Once the transition period is past, overall average annual profits increase significantly — they averaged $10,000 more per year after transition ($32,083 vs. $22,027) in this study. That's *without* government program support. Factor that in, and post-transition profits jump to an annual average of $44,202, compared with just $17,790 during transition.

❖ In the conventional scenario, average annual profits (despite government programs) *declined*, from $36,125 during '81 to '84, to $34,738 during the next five years. Without deficiency payments when grain prices fell in the mid-'80s, many conventional farmers would have gone bankrupt trying to grow just corn and soybeans.

❖ The "economic transition" was at least twice as long as the biological transition in this study. Although it took only four years for soil health to improve dramatically during the FST trial, researchers estimated that it would take a farmer an additional five years to fully recover his transition costs.

That last finding is pretty scary, isn't it? After all, you have to feed your family *during* your farm's transition period, while still being able to afford equipment replacements and other capital improvements. But the up side here is that for the purposes of this financial study, researchers assumed the low-input farmer had a *difficult* transition. They purposely increased the farmer's equipment- and repair expenses and downgraded hay and straw prices, assuming that an inexperienced low-input farmer might have done a poor job of managing forage crops.

Capital-replacement costs, the researchers note, could be lowered significantly with timely repairs and maintenance. Labor cost estimates could be reduced if additional family help were available. Moreover, maximizing farm profitability hasn't been the main objective of the FST project. Rodale Institute researchers wanted a better understanding of the transition period and soil biology. Working farmers might have chosen different grain crops during some years, based on market outlook and weather conditions.

Cross-compliance regulations in government programs have hampered the adoption of better rotations by low-input farmers, the researchers point out. Relaxing program regulations (beyond the minor changes found in the 1990 Farm Bill) could make it easier for the government to *support* transition, rather than hinder it.

One other point to note: FST compares two farming extremes. If you reduce your use of farm chemicals gradually — while making use of livestock manure and/or legume-based fertility — you could see a quicker recovery of transition costs than you would from an overnight elimination of synthetic inputs.

---

[1]Hanson, James C. *et al*, "The Profitability of Sustainable Agriculture on a Representative Grain Farm in the Mid-Atlantic Region, 1981-89," *Northeastern Journal of Agricultural and Resource Economics*, Vol. 19, No. 2 (October 1990), pp. 90-98.

Dr. Bill Liebhardt, director of the Sustainable Agriculture Research and Education Program at the University of California and former RIRC director. "If your 5- or 10-year yield average for corn is 120 bushels per acre, then fertilize accordingly. Don't set a yield goal of 150," he says.

"You can make the transition without new equipment and without additional inputs. You have to learn to use what you have," Liebhardt adds. To ensure that you're not overfertilizing, get a second opinion on soil tests, he suggests. In many cases, you can cut or eliminate routine P or K additions without hurting your bottom line.

It Gets Easier

In '86, FST researchers shifted the study's emphasis from transition feasibility to long-term reliability. Besides continuing to look at how crops performed in the different systems, they wanted a better understanding of the soil biology and economic risk of each system.

Their working assumption: Pesticide-free, low-input systems reduce plant stress and increase biological efficiency, compared with conventional cropping systems. So low-input systems should show equal or better plant growth and yield compared with a conventional system, while conserving or enhancing the resource base.

Through collaboration with other researchers from land-grant universities — Michigan State, Cornell, Ohio State, Penn State and elsewhere — plus the USDA Agricultural Research Service, RIRC staff are learning even more from the living laboratory that the FST field provides. The results are helping farmers increase soil tilth and soil life, make better use of nutrients in their soil, harness the power of legume nitrogen and minimize weed, disease and insect problems.

"You can make the transition without new equipment and without additional inputs. You have to learn to use what you have."

What are some of the discoveries FST researchers have made in the past few years? And how can you and your farm benefit from these findings?

The FST data show that by using animal and/or green manures, well-planned and diverse crop rotations, timely field operations and nearly continuous live plant cover — while avoiding synthetic fertilizers and pesticides — low-input systems:

- ❖ Increase soil water-holding capacity, allowing crops to better withstand moisture extremes.
- ❖ Exhibit higher microbial activity, enhancing soil health.
- ❖ Show greater total soil carbon and nitrogen levels than typical conventional systems, and greater mineralization rates, thereby improving nutrient uptake.
- ❖ Increase soil organic matter, especially the biologically active fraction, which increases the amount of available nutrients.
- ❖ Allow for greater water infiltration, reducing soil erosion and increasing water-use efficiency.
- ❖ Lower input costs while maintaining or improving yields.
- ❖ Let you almost ignore weeds in many fields without hurting profits.
- ❖ Reduce financial risk after the transition period.

During the drought year of 1988, the FST's low-input corn following a red clover cover crop outyielded conventional corn following corn by 24 bushels per acre, and conventional corn after soybeans by 5 bushels. This was a direct result of improved soil water-holding capacity in the low-input plots, says Janke.

RIRC researchers emphasize the value of including green manures in your crop rotation to provide as much year-round cover as possible. "Any green-manure cover suitable for your area will provide balanced fertility and improved soil structure, as well as give your crop a competitive advantage over weeds," says Peters.

"A big flush of nitrogen, as often occurs with commercial fertilizer or raw manure, will tend to encourage rapid weed growth (and reliance on herbicides). A slower nitrogen release, whether from a legume green manure or composted manure, benefits the crop *and* the soil, our data show, while diminishing the need for herbicides," he says.

"For corn that follows a legume, we delay spring planting until mid-May (more than two weeks past the conventional corn planting date in our area) to allow the cover crop to put on as much biomass as possible. In just a few days between late April and early May, the amount of cover crop biomass can double, improving soil fertility and weed control.

The later-planted corn quickly catches up to the conventional plots even without any starter fertilizer, thanks in part to warmer soil conditions."

RIRC staff overseed red clover and orchardgrass into wheat in early spring in LIA, the low-input system that also uses cattle manure. They drill hairy vetch after wheat harvest in LICG.

"We use a seeding rate of 10 pounds per acre for red clover, which provides 120 to 150 pounds of total nitrogen at plowdown," says Peters. "In the first year after incorporation of clover, we estimate [from FST nitrogen-tracking studies] about 20 percent of the legume nitrogen is readily available to the crop, supplementing N reserves from soil organic matter." They also seed 5 pounds of orchardgrass with the red clover, to increase the green-manure biomass, provide a more-balanced forage and speed drying of hay in the field.

During the trial's first 10 years, the researchers relied on a mix of red clover and alfalfa for legume N before corn in LICG. Now they seed hairy vetch at 32 pounds per acre. "We found that vetch seems to give quicker improvement in soil properties such as tilth and nitrogen availability. There are tradeoffs. Vetch produces more nitrogen than clover does, but is more expensive to seed," says Peters. "We drill vetch a month after wheat comes off, so we have to till weeds under, cultipack, then plant. You can just frost seed clover in spring onto existing wheat fields. But we want to be sure we get enough N into the system. Clover runs a fine line there."

They've seen mixed success when trying to establish a cover crop in grain corn to provide winter soil protection and weed suppression. "The large amount of corn residue remaining after grain harvest suppresses growth of an overseeded cover crop," he notes. "We've experimented with rye, ryegrass and various legumes, broadcast in June at last cultivation or in late summer with a high-boy seeder. We like overseeding annual ryegrass the best, usually at lay-by when corn is 12 to 15 inches high.

"Overseeding soybeans during leaf yellowing in early September also has been successful. We usually overseed winter rye in LIA soybeans now, although hairy vetch and crimson clover work well too. In LICG, we follow soybeans with drilled winter wheat."

Plan Year-Round Cover

From '86 to '90, RIRC staff found they often could increase net return in LICG by drilling soybeans into standing small grains before boot stage. But they weren't happy with the reliability of relay cropping.

"For the first five years, we were basically taking off one cash crop per year and our weed-control costs were cultivation," observes Janke. "Relay cropping gave us four cash crops in three years. Plus, we weren't cultivating soybeans anymore, because the small grain in the relay-crop system suppressed weeds early, and the soybean canopy suppressed them late."

RIRC scientists first tried relay cropping in FST in '86. In late March, they drilled spring barley in 7-inch rows at 2 bushels per acre. In mid-May they drilled 1.5 bushels of soybeans per acre into the foot-high, tillering barley. No additional field operations followed soybean planting other than small-grain harvest in July and soybean harvest in October.

The barley yielded 30 bushels per acre. That's 15

RIRC Photo

For winter cover in Low-Input Cash-Grain plots, RIRC staff prefer to overseed ryegrass into corn at lay-by, using a Cyclone seeder and a Buffalo cultivator.

Focus On *Your* Farm

Site-specific factors, operator experience and climate are important considerations when cutting your chemical use, says Dr. Jim Tjepkema, Rodale Institute Midwest Network coordinator. "Soil types, field fertility, weed levels, crop budget and local weather patterns — plus your farming philosophy and willingness to experiment with alternative methods — can influence your selection of cover crops, seeding rates, planting dates, sidedress rates and weed-control methods," he says.

Working closely with more than a dozen network co-operators each year, Tjepkema says he has seen the need to proceed slowly when trying a new farming method or improving a rotation. "A new crop or an unfamiliar technique can surprise you by performing extremely well in a given year, or shock you with poor results when something unexpected happens." But that's true regardless of what farming system you're using, he adds.

"When eliminating herbicides, for example, you often find that weed control is easier than you expected, but can be a problem initially in some fields. You can have surprises *with* herbicides too, though, such as poor incorporation or carryover effects," notes Tjepkema.

"At the Rodale Institute Research Center, there are tradeoffs we can tinker with over time that low-input farmers must address *each year* to maintain profits," says agronomist Steve Peters, project leader for the Institute's Farming Systems Trial (FST). Target plant populations, crop and variety selection, and choice of tillage are some of the decisions farmers face regularly, he notes. For transition studies, researchers typically keep many of these factors constant, to make it easier to compare different farming systems.

Not every new method or specialty crop works the way you expect it to.

"We see FST as an idea generator for farmers, as well as an ongoing study of soil dynamics," says Peters. Many of the Institute's farmer-cooperators have tried techniques tested in the FST fields near Kutztown, Pa. — or have suggested alternative methods based on their experience, says Tjepkema. The common denominator: They're all seeking ways to increase their use of *on-farm* resources.

By including hairy vetch in his rotation, for example, Terry Holsapple, Greenup, Ill., found weed levels diminishing in his fields, and the pool of available nitrogen increasing. "It's easy to get 200 to 250 pounds of N per acre from a vetch crop around here," Holsapple says.

"You have to find a legume that works well in your area, though," says Carmen Fernholz, Madison, Minn. He's discovered that some vetches fail to survive cold Minnesota winters. Fernholz also tried interseeding vetch or other legumes with small grains in spring, hoping to leave the covers in for a year. But when he harvested the grain, the legumes often weren't vigorous enough to compete with weeds in some of his fields. Now, he prefers to plant winter-hardy legumes with an old endwheel drill behind his chisel plow when tilling grain stubble in fall.

"Farmers won't adopt covers if it costs too much," says Boone, Iowa, farmer Dick Thompson. To minimize seeding costs, he's adapting use of rye/vetch covers drilled on ridgetops rather than broadcast. There's still lots to learn about ridge tilling, he says, but you can sure lower fuel bills while building organic matter *and* controlling weeds on ridges.

Rich Bennett, Napoleon, Ohio, hopes to use hairy vetch after wheat, to supply nitrogen for corn and to help eliminate use of a Lasso-atrazine banding. "A late-spring nitrate test showed high N levels when I first tried hairy vetch in '91. In '92, though, when I didn't incorporate the vetch before no tilling corn, the test showed I wasn't getting much legume N. So I went back in and put on 90 pounds of sidedress N per acre," Bennett says.

"We're not sure yet how well hairy vetch performs in a no-till system such as Rich Bennett's, where he kills the vetch with a herbicide before no tilling corn," says Tjepkema. "Vetch doesn't provide as thick a root system as many other covers do. And with his sandy soil, Rich may find that vetch isn't as consistent a performer as it is for Terry Holsapple."

Tom Culp, Lexington, Ohio, sees good results from aerially seeding a rye/vetch mix into soybeans two weeks before leaf-yellow in mid-August. "We combine beans and let vetch and rye grow over winter, then no till corn after killing the vetch/rye with herbicide. The covers work well on our rolling silt-loam fields."

Not every new method or specialty crop works the way you expect it to. Culp had experimented with growing triticale, a wheat-rye cross, as a low-cost alternative to raising corn grain for his hog ration. He stopped growing it, though, because of inconsistent yields and disease problems. "I tried an improved variety in '91, but I couldn't plant it early enough because of a wet spring," says Culp.

Barry and Dean Eppley, Wabash, Ind., had early success with growing 60 acres of clear-hilum soybeans for the tofu market. But wet weather and weedy fields have hampered their efforts to grow other specialty crops such as blue corn, adzuki beans and black soybeans on 90 certified-organic acres. "We can't justify larger acreages of organic crops — yet," says Barry, who's convinced the profit potential is there.

"Each year's different on your farm," says Tjepkema. "Be flexible."

percent below normal, because the combine head was kept high to avoid injuring the beans. Field traffic when drilling soybeans also hurt some of the barley. But the small grain paid for itself and helped suppress weeds in the 48-bushel soybean crop.

The barley-beans combination netted $148 per acre, compared with $90 per acre for the 42-bushel conventional soybeans that year. Similar results were seen with winter wheat and soybeans in '87 (32 bushels per acre and 44 bushels per acre, respectively), and barley and beans again in '90. But weather extremes in '88 and '89 caused problems. Lack of moisture in '88 meant that spring barley used up too much soil moisture, causing a soybean crop failure. Barley yielded well, however, at 61 bushels per acre.

In '89, wet weather caused soybeans to grow above the barley canopy. That lowered barley grain yields, which were about 8 bushels per acre.

"All told, we had better results growing soybeans with winter wheat than with spring barley or oats," says Peters. But relay cropping seemed risky. "So we changed LICG to a three-year rotation of soybeans, then winter wheat, followed by a late-summer seeding of hairy vetch instead of red clover, and then corn with ryegrass overseeding. That gives us almost year-round cover."

RIRC Photo

Spring barley shades out weeds in soybeans and boosts return, but moisture extremes make relay cropping risky.

Will weeds be a problem without herbicides? "Weeds are almost always more abundant in our low-input fields than in the conventional ones," says Peters. But weeds have rarely been an economic factor since transition, he notes. "The higher weed levels have caused a yield reduction in only two of the 14 low-input treatments since 1986. In one of these, a short-season corn variety produced a later canopy cover and enabled more weeds to take hold. In the other case, drought reduced the corn yield."

Weeds lowered soybean yields in only two of the nine low-input treatments since '86, adds Janke, who established comparison subplots kept weed-free by handpulling. Weed levels have typically been highest in LIA plots, she says. "This could be due to introduction of weed seeds when researchers add semi-raw cattle manure to the plots, or it could be due to regrowth of weeds in the hay that preceded corn."

"We see year-to-year shifts in weed species, as you would expect in most fields," says Peters. "Time of cultivation, varying moisture levels, crop sequence and other factors affect weed populations. Effective long-term weed management, we've seen, requires the use of cultural controls such as crop rotations, cover cropping and, where suitable, intercropping or relay cropping.

"Weeds aren't all bad, I might add. They loosen the soil, reduce erosion from rain runoff and keep nutrients in the system for the following year's crop. You have to determine what weed levels you can permit without sacrificing yield.

"Economic weed threshold levels are usually a lot higher than visual comfort levels. In FST, we've probably had *more* weeds year to year than most low-input farmers would. That's because they'll handpull weeds in trouble spots, do extra cultivation in problem fields or make other management decisions that a replicated study such as FST won't allow."

When converting to a low-input system, you might not want to eliminate herbicides in one season, Peters points outs. "If a field has been pretty productive, and soil fertility and structure are good, then you probably can drop herbicides altogether. If it's a very weedy field or has low fertility, go a little slower until you're happy with your yields. In row crops, first do some testing, or banding or spot spraying after cultivation. Your progress will depend on your cultivating skills, crop rotation and field conditions.

"We plan to continue FST as long as it provides helpful information on pesticide-free cropping systems as compared with conventional practices. Along with our many collaborating researchers, we're focusing on soil biology, soil health and long-term reliability of the three cropping systems. We also hope to do some feed-quality research. The study gets more interesting every year."

Reducing Tillage *Without* Chemicals

To control erosion and to meet conservation compliance provisions of the '85 Farm Bill, many of you have been reducing tillage in the past few years. No-till acreage in the United States will jump about 13 percent in '93, to nearly 32 million acres, according to a tillage-practices survey conducted annually by the editors of *No-Till Farmer*. Ridge-till acreage will increase about 3 percent, to 3.5 million acres, they estimate.

Unfortunately, many farmers apply *more* herbicide to control weeds when they reduce tillage. Not happy with that — or with their own reliance on moldboard plowing to incorporate legumes in FST — RIRC researchers decided to see if it's possible to reduce tillage profitably *without* chemicals. "We want to develop methods of doing this without hurting yield," says Peggy Wagoner, project leader for the Low-Input Reduced Tillage (LIRT) trial started in 1987.

New Farm *Photo by Ed Landrock*

Preparing to harvest soybeans in LIRT plots in fall '92, RIRC farm manager Jeff Moyer and field foreman Owen Maguire check details with Peggy Wagoner, LIRT project leader, and Lucia Longnecker, agronomy technician.

The study compares four tillage methods: moldboard plowing, chisel-disk (chisel plowing then disking, leaving some residue), ridge-till and no-till, plus a mixed-tillage regime. When RIRC staff have established good legume stands and weather conditions have been adequate, they've generally had excellent yields in following crops of corn, soybeans and small grains, even with reduced tillage. They've done this *without* herbicides, by killing cover crops with a flail mower and by controlling weeds with two rotary hoeings and two cultivations in all but strict no-till plots.

They've had mixed success with some of the no-till and ridge-till treatments, however, and are continuing to hone low-input systems that work well with reduced tillage. When covers failed or grew poorly, for example, crop yields have generally been lower than for similar tillage treatments where conventional herbicides and fertilizers have been used. When they overseeded covers, the researchers saw poor stands in some years because of germination problems related to high-residue conditions.

"We found that overseeding hairy vetch is risky. We need to drill the vetch in," says Wagoner. "When we overseeded hairy vetch into relay-cropped soybeans at leafdrop, the vetch seeds would land primarily on old barley stubble, providing poor seed-to-soil contact. Plus the high residue was a good habitat for slugs that damaged the young plants."

"If you're reducing tillage, you're going to have to select equipment suited for high-residue situations and harder soils," says Jeff Moyer, RIRC farm manager. "You can't expect to cultivate with an old multitined cultivator. You'll need a heavy-duty one, as well as something to

Low-Input Reduced Tillage Rotation

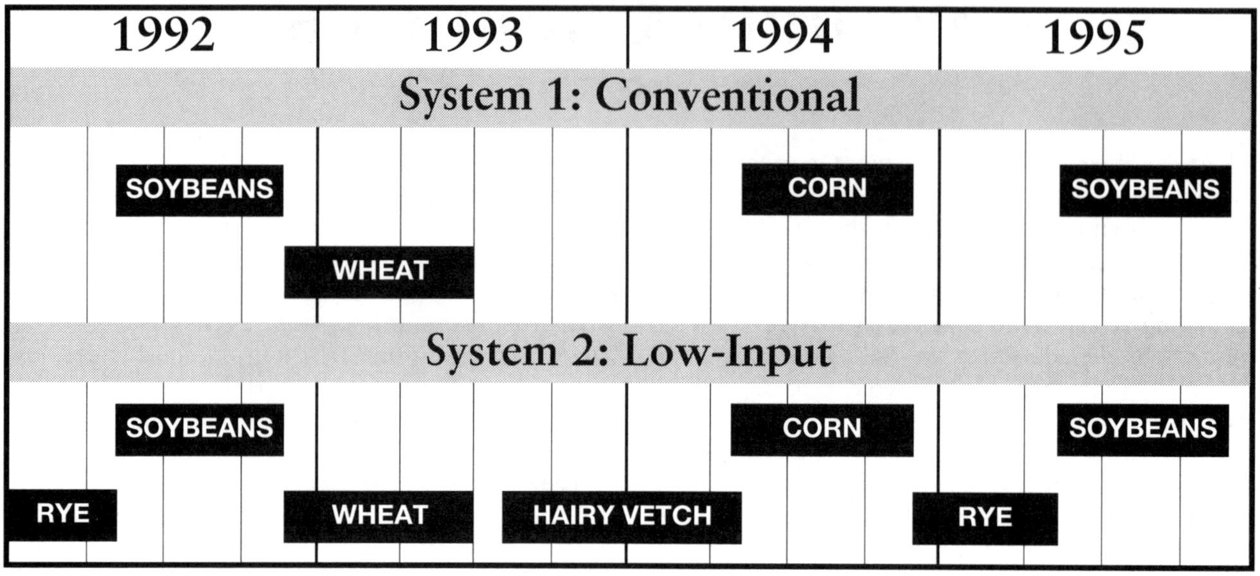

handle the residue. That goes for your rotary hoe as well, if you're using a hoe.

"In the low-input plots, we count on cover crops for canopy to reduce weeds, for nitrogen and for organic matter," Moyer continues. "So with a poor cover, we weren't able to get adequate crop growth. The trial design prevents us from changing the crop selection in a given year, but we can do additional cultivation if we choose. We want to keep the number of cultivations the same, however, so we can compare data more directly."

"We try to time mechanical cultivation so it provides good weed control," Wagoner says. They typically rotary hoe twice a half-inch deep at 8 to 10 mph within two weeks of planting corn or soybeans. They cultivate corn twice in the next three weeks, with a Buffalo 4600 high-residue cultivator, 2 inches deep at about 5 to 6 mph. "We cultivate slower at times, depending on the crop stand and weediness," says field foreman Owen Maguire. "For soybeans, we cultivate at about 3 to 5 mph, and do second cultivation just prior to canopy closure."

"We haven't done any extra cultivation or skipped a hoeing so far in LIRT. A farmer could," Wagoner points out.

Sidedressing nitrogen to make up for a failed legume cover is another option farmers have, says Moyer. "Bagged and liquid fertilizer, animal manure or compost — farmers have lots of choices," he notes.

Even when a cover fails in a low-input treatment, however, carryover nitrogen from a *previous* stand can supply some fertility, the researchers observe. In '91, for example, a moldboard-plowed treatment produced 102-bushel corn despite failure of a preceding vetch stand. "The corn may have benefited from vetch-supplied nitrogen carried over from 1989," says Lucia Longnecker, RIRC agronomy technician. In '89, vetch provided 260 pounds of N per acre in that treatment, she says.

Mow-Kill Controls Covers

When covers grow vigorously, how do you keep them from getting away from you without herbicides? "When killing a cover [mechanically], the type of mower isn't as critical as the timing of the mowing. One mower might chop finer or distribute better than another, which may be important for small research plots, but isn't as critical for a farmer. Timing *is*," Moyer says.

"We time the mowing so that it will kill most of the cover. It's crop-specific, based on the cover and how it's doing in the season. In general, you mow-kill an annual or winter-annual cover when it has flowered but before the seed matures.

"Unfortunately, many covers don't mature as uniformly as you would like. Roundup will kill covers, of course, and plowing will too, but mow-kill and planting into cover residue means you have to time it right to prevent regrowth or the cover will act as a weed.

"If you want to delay cover killing, you can plant *into* it — say with corn — and kill the cover later. The risk is that a corn crop will get too tall before you actually mow the cover, so you need to keep checking the corn. Farmers also may choose to work with oats or other crops that winterkill. That gives a nice mat you can plant into."

"There's still a question as to when exactly is the best time to mow hairy vetch," says Wagoner. "You have to be sure it's flowering and has some seed pods starting to form. If you can mow it when you have dry weather, that also helps kill it, as a nice rain right after mowing will cause some regrowth."

In low-input plots, LIRT researchers plant hairy vetch about a month after winter-wheat harvest. They use either a John Deere grain drill or a Tye no-till drill to seed 30 to 40 pounds of vetch per acre. The legume overwinters, then puts on spring growth. "We usually mow-kill hairy vetch four to seven days after planting corn into the vetch in early May," says Wagoner.

After corn harvest, they seed rye at 1.5 bushels per acre and mow it in mid- to late May when it's fully headed. "We don't see 100-percent kill from a single mowing, but cultivation of the following soybean crop takes out most of the remaining rye," she says.

New Farm *Photo by Ed Landrock*

Agronomy technician Lou Saporito measures the water-infiltration rate in corn grown in a reduced-tillage plot at the Rodale Institute Research Center. Ongoing trials show that low-input farming methods improve the efficiency of water use by crops and reduce erosion.

Besides its allelopathic (natural weed-suppressing) ability, rye fits the rotation well. "We plant rye fairly late, after corn harvest in late October or early November. It puts on good growth even in cool fall temperatures," says Wagoner.

Varying Tillage Helps Too

The "wildcard" mixed-tillage regime, in which primary tillage is determined yearly, has provided RIRC staff with exceptional results during the trial's first five years. "Mixed tillage has been our shining star," says Moyer. "Anytime you can maximize your options to suit the weather, weed levels and crop options, that's the way to go.

"For our no-till treatment, we *have* to no-till. But with other tillage tools available, a farmer could decide to chisel plow or moldboard, depending on soil condition, weather, and crop choice."

"In the mixed-tillage treatment, we've always been able to no till the primary crop as long as we've gotten a good cover," says Wagoner. "In '89, for example, we no tilled corn into an excellent stand of hairy vetch that had been planted the previous fall after moldboard plowing. At 140 bushels per acre, corn yields were comparable with the conventional corn treatments, and at an equal or higher return per acre."

Cash returns averaged $220 per acre for the mixed-till treatment as well as for the conventional moldboard-plow treatment, according to an economic study by Dr. Dave Watt and Bruce Dahl of the North Dakota State University Agricultural Economics Department. The other conventional treatments returned from $120 to $180 per acre.

That same year, low-input corn in the strict no-till treatment yielded a dismal 63 bushels per acre. Poor establishment of an overseeded cover in '88 led to a nitrogen deficiency in '89, notes Wagoner.

In '91, RIRC staff chisel plowed the mixed-tillage plots following a poor vetch stand and still managed to harvest 97-bushel corn. "That wasn't tops for us, but it was a good yield," she says. That year, the conventional chisel-disk treatment yielded 120 bushels per acre, while a low-input chisel-disk treatment gave 130 bushels.

During LIRT's first four years, the researchers compared the tillage regimes under three cropping systems:

❖ A **conventional corn-soybean rotation** relying on Penn State Extension recommendations for fertilizer and pesticides.

❖ A **low-input corn-soybean rotation** using no

So What About Weeds?

If you expect weed pressure to increase when you eliminate herbicides, you could be surprised. With adequate cultivation and a diverse rotation that includes cover crops, your weed worries should diminish over time — even if you reduce tillage.

That's what Rodale Institute researchers have seen in their Low-Input Reduced-Tillage trial (LIRT), begun in 1987. The long-term study compares the effects of different tillage regimes on crop performance, soil organic matter, infiltration rates and other factors, including weed levels.

In low-input, high-residue plots where hairy vetch cover crops have established poorly, populations of annual grasses and ragweed *have* tended to increase, reports agronomy technician Lucia Longnecker. That has hurt yields in some of the low-input plots, particularly ridge-till and no-till treatments where corn and soybeans were being rotated.

By including small grains such as wheat, oats or rye in the rotation, however, the researchers found that biomass levels of ragweed, lambsquarters, and other annual weeds dropped dramatically.

"Problem areas of perennial weeds are also under control — they haven't enlarged significantly, although we've seen some increase in thistles," says Longnecker.

To determine if weeds have an impact on yields, the researchers employ "plus/minus" subplots in each treatment plot, based on a research design suggested by Dr. Rhonda Janke, director of the Rodale Institute Research Center. They handweed 10- by 15-foot subplots throughout the year, then compare crop yields with adjacent subplots that weren't weeded.

In '91, the subplots showed low-input no-till treatments suffered heavy weed competition, says Longnecker. Hairy vetch covers failed to establish in those plots, due to high-residue conditions and slug damage. With negligible N from the legume, the corn crops were nitrogen-deficient and couldn't grow, she says. "Even the weeds in our no-till plots looked poor that year," says Jeff Moyer, RIRC farm manager.

A low-input ridge-till treatment also showed losses from weeds and a poor vetch stand, and yielded a meager 27 bushels of corn per acre. The ridge-till treatment with a stronger vetch stand and more diverse rotation yielded 101 bushels per acre despite moderate weed pressure. With the same rotation, chisel-disk and moldboard plots yielded 130 and 124 bushels per acre. Those plots had excellent vetch stands and weeds didn't lower yields.

The LIRT data confirm that weed biomass levels are better indicators than weed counts when deciding if weed pressure will hurt yields, observes Janke. "Just because you see some weed seedlings doesn't mean you're going to have a lot of weeds later in the season. In a given year, weed flushes in May decline by July in the low-input plots." Cultivation and canopy closure kill many of the weeds, she notes.

"Although weed biomass levels peak in early summer, they decline by September. You should be concerned [only] if you have many weeds remaining after last cultivation. Having a lot of weed biomass at silking in July will cause corn yields to suffer."

In many cases, it might not be worth going in with a post-emerge herbicide, says Peggy Wagoner, LIRT project leader. "In the conventional cropping system for a corn-soybean rotation [using Extension-recommended pesticides and fertilizers], we've had varying success with herbicides, especially in dry years," she points out. "And in '90 (a fairly normal year for rainfall), the recommended herbicides failed to control ragweed in the conventional plots, hurting soybean yields.

"In '92 (a wet year), we had to spend a lot of money on herbicide to get good weed control in soybeans in the conventional cropping system. We used 1.5 pints of Dual and 10 ounces of Gemini per acre at planting. Then as we saw ragweed coming in, we added Classic at two-thirds of a pint per acre in the moldboard and chisel-disk treatments. In the no-till plots, we had to come in later with 1 pint of Poast and 2 pints of Booster E per acre to control foxtail.

"We got by with just banding some pre-emerge over the planted row in the ridge-till treatment and cultivating to control weeds prior to soybean canopy closure."

> ***Weed biomass levels are better indicators than weed counts when deciding if weed pressure will hurt yields.***

"Farmers can rely on experience and observation to figure out their own weed thresholds," says Longnecker. If you have time to experiment and want to see if weeds are actually hurting your yields, consider estimating weed biomass levels in some of your fields, suggests Wagoner.

Here's how: Assuming decent fertility and adequate moisture, a rule of thumb is that weed dry matter exceeding 500 pounds per acre will hurt yields. That's roughly equivalent to 1 pound of (oven-dried) weeds per 30 row-feet if you're on 36-inch rows, or per 35 row-feet on 30-inch rows. When harvesting weeds, be sure to snip them at the soil surface and include weeds to midrow on each side of the row. Dry the weeds in a standard oven at low temperature (120 F) and weigh.

Rather than spending time *measuring* your weed pressure, however, why not invest the time *preventing* weeds in the first place? Besides including small grains in your rotation, you can also control weeds with timely cultivation and a mixed-tillage regime, notes Wagoner. "We've seen very good weed control and great yields by varying the primary tillage method and by ensuring good seed-to-soil contact for all crops."

RIRC Photo by Peggy Wagoner

As part of Rodale Institute's reduced-tillage trial, field foreman Owen Maguire chisel plows 8 inches deep to kill hairy vetch in a low-input plot. Corn planted two weeks later yielded 130 bushels per acre.

fertilizer or pesticides, and cover crops for nitrogen and weed suppression.

❖ A **three-year, low-input diverse rotation** — corn, wheat with double-cropped soybeans, a mix of oats and berseem clover, then hairy vetch — using minimal fertilizer and no pesticides.

The staff gathered data from 13 treatments (four tillage methods times three systems plus the mixed-till treatment) replicated six times in a well-drained, hilltop field of shaley silt loam. The field had been farmed organically in previous years with field crops, and organic-matter levels ranged from 3.1 to 3.7 percent.

Some of their conclusions during this period (in addition to those already noted) include:

❖ Good seed-to-soil contact is important for establishing both grain and cover-crop stands.

❖ Higher levels of tillage tend to provide better seed-to-soil contact and better establishment of crop stands.

❖ Large-seeded crops such as corn and soybeans appear to tolerate reduced-tillage regimes better than small-seeded cover crops.

❖ Weed control and good crop establishment are harder to achieve in ridge-till and no-till systems, compared with moldboard-plow and chisel-disk regimes.

❖ Rotation of tillage regimes can provide good economic yields while reducing the need for herbicides and chemical fertilizers.

As noted previously, the researchers weren't happy with the reliability of overseeding in many of the low-input treatments. They also found they were spending more time than they expected to gather and interpret weed data. (See sidebar, "So What About Weeds?")

In fall '91, they changed all the treatments to a three-year rotation of corn, soybeans and wheat. "In the low-input plots now, we drill hairy vetch after wheat to supply nitrogen for corn, and drill rye after

Plan Ahead to Reduce Tillage *and* Chemicals

To decrease weed pressure with reduced tillage, you'll need some experience with low-input methods. Here are a few general tips:

❖ Choose cover crops appropriate for your region and for your primary crops, as well as for your equipment.

❖ Test, on a small scale, any cover crops you haven't grown previously, before making large plantings.

❖ Plan to maintain some soil cover — living or dead — for the maximum amount of time.

❖ Include small grains in the rotation to provide a midsummer tillage option after harvest of small grains.

❖ Consider a mixed-tillage regime. If you're a no-till farmer, you may want to plow at times to help break a weed cycle and to reduce herbicide use.

❖ Hone your rotation and cultivation skills through on-farm trials comparing reduced-herbicide and herbicide-free plots.

corn for weed control before soybeans. That makes it more of a low-input tillage comparison rather than a cropping-systems experiment," says Longnecker.

"Winter wheat helps us alternate warm-season and cool-season crops," Wagoner points out. Based on the first four years' results, "we do a better job of controlling weeds with a diverse rotation that includes small grains and excludes relay cropping.

"We're also looking closer at *rotation* of tillage, by alternating chisel disk and no till in two of the low-input treatments. In one, we do the major chisel tillage prior to putting in a small-seeded crop such as wheat or a rye cover, then no till the following crop (corn or soybeans). In the other, we chisel just before a large-seeded crop such as corn or soybeans, then no till rye or wheat. So we're comparing chiseling to no tilling the main crop. In these treatments, we *always* chisel before planting hairy vetch, because it is critical to get a good stand of vetch for the succeeding corn crop."

It's hard to rotate a ridge-till treatment with another primary tillage method, of course. But RIRC staff also are comparing a reduced-residue ridge-till treatment with a traditional ridge-till treatment, Wagoner adds. "We're seeding a cover just on the tops of ridges for a reduced-cover ridge-till treatment. We're comparing that with drilling a cover over the entire plot."

They're also studying the value of cultivation in low-input no-till systems. "We're no tilling soybeans on 30-inch rows to allow cultivation, and comparing that to another no-till treatment on 8-inch rows without cultivation," says Wagoner. "Cultivation will help control weeds and stimulate release of nitrogen in the soil."

She says they're encouraged by what they've seen so far in LIRT, especially in the mixed-tillage treatments. "When you not only rotate crops but also rotate tillage, that gives the option of disrupting the soil with plowing to better establish a cover and help break a weed cycle. In following years, you could use low tillage — and still have good yields *without* chemical herbicides or fertilizers."

RIRC staff plan to continue LIRT for 10 to 20 years and to modify the treatments periodically, says Wagoner. "We're learning more about what does and doesn't work in our efforts to develop low-input reduced-tillage cropping systems."

For more information, or to join hundreds of other farmers who tour the research fields annually, contact: Rodale Institute Research Center, 611 Siegfriedale Road, Kutztown PA 19530, (215) 683-6383.

New Farm *Photo by Ed Landrock*

In '92, a low-input no-till treatment with mechanical cultivation yielded 37-bushel soybeans, RIRC scientists Peggy Wagoner and Lucia Longnecker found. "That's not bad for plots no tilled for five years without any herbicide," says Wagoner.

Chapter 6:
'The Last Chapter Hasn't Been Written'

BOONE, Iowa — Half a dozen hay racks are overflowing with curious farmers, scientists and lawmakers. Dick and Sharon Thompson's annual field day is ready to begin. If you count the contingent from the People's Republic of China, some have come literally halfway around the world to visit this pioneering regenerative farm.

A TV camera crew scrambles for a better angle, and the bullhorn crackles to life as Dick welcomes the crowd. "We don't have all the answers," he proclaims. "We're always looking for something better, so we reserve the right to change our minds. The last chapter hasn't been written."

Some visitors may be disappointed by the caution in Dick's opening remarks. But their concern will be short-lived. In about two hours, when the convoy of hay racks returns from touring the 300-acre farm, clusters of folks will be discussing passionately how the Thompsons:

❖ Grow 150-bushel corn without herbicides for $45 less per acre than conventional farmers.

❖ Produce 50- to 60-bushel soybeans for $48 less.

❖ Plant cover crops to protect soil, suppress weeds and reduce purchased N.

Dick and Sharon Thompson
Boone, Iowa

Started cutting inputs: 1967
100 acres corn
60 acres soybeans
60 acres oats
40 acres grass/legume hay
40 acres pasture
300 acres

Livestock: 80-cow cow-calf herd
70-sow farrow-to-finish operation

Growing Season: 158 frost-free days
32 inches of precipitation

❖ Raise healthy cattle and hogs without antibiotics.

❖ Store and handle manure to make the most of its nutrients.

❖ Plant more than 200 research plots that satisfy the needs of both researchers and farmers, while helping the Thompsons fine-tune their own management methods.

The heart of the Thompsons' message is that farmers can adapt — not adopt — their techniques. Test new practices on your own farm and let nature teach you how to improve your bottom line, they stress. "There are no simple recipes," says Dick. "Farmers have got to think for themselves instead of letting everyone else do their thinking for them."

The Thompsons have been thinking for themselves for more than a quarter-century now. And while they've changed their production methods often during that period, they have never changed their original goal: "Profitable and environmentally sound farming — pure and simple," says Sharon. "It's got to sustain the land, the soil, the people, the communities and the pocketbook. It all has to fit together."

Whipping Weeds Naturally

Several practices work together to help the Thompsons grow ridge-till corn and beans without herbicides. Weed control begins in fall, when the Thompsons drill weed-suppressing cover crops (more about these later). In spring, they spread about 15 tons of manure-sludge mix per acre just ahead of the planter.

Their modified Buffalo ridge-till planter skims the top 2 inches of soil off the ridge, taking the cover crop, weed seed and manure out of the row. Soil from the ridge covers the manure, helping to reduce N volatilization. "You can lose a quarter to a third of your N in 36 hours if you don't incorporate the manure," observes Dick. The only drawback is that spreading slows down planting.

"The Buffalo is the only brand of planter we could modify to move the depth gauge wheels off the ridgetop," he continues. He removed the coulter, split the center depth band and mounted the two sections 16 inches apart. "This reduces compaction and disturbance on the ridgetop. That's especially important for controlling foxtail, because it germinates faster in a tight soil."

The weed-seed deflector on the Thompsons' planter is tight behind the 14-inch sweep to make sure all weed seeds are moved out of the row area. The trailing residue guards are sharpened and point down and out to make sure weed seed and residue continue to flow away from the row far enough that the cover disks don't pull them back in. The Thompsons use 36-inch rows so they have enough room to move residue. "You can build a taller ridge, so it's still higher than the row middles after planting," says Dick.

The 1-inch-wide press wheels firm soil only along

the seed row. The rest of the ridge remains unpacked, to permit easy root growth. Cover disks are set to mound about 2.5 inches of loose soil over the seed. When this soil dries out, weeds won't germinate. "I can rotary hoe this mound of soil without disturbing crop seeds," observes Dick.

Thompson customized his Buffalo ridge-till planter to minimize compaction and disturbance that he says "wake up" weeds.

The Thompsons are testing different starter fertilizers, placing them 2 inches below the seed, where they'll give corn a quick boost instead of feeding weeds. "And no spring tillage on the ridges," stresses Dick. "All that does is wake up weeds in the row."

In '89 and '90 test plots, the Thompsons tried pre-cultivating cover crops before planting, either to kill rye in the valleys between ridges or to speed N release from hairy vetch. In both cases, weed pressure increased dramatically. "Pre-cultivation to remove cover crops in spring is not a good practice," says Dick.

High plant populations help shade weeds. The Thompsons plant 12 soybeans per foot (175,000 seeds per acre) and corn at 6-inch spacings (29,000 seeds per acre) in 36-inch rows. They choose varieties that emerge quickly, have good early root growth and perform well at high populations.

With no herbicides, weed control in the row is the toughest job. "The row area must be kept clean," stresses Dick. In the past, he normally made a double pre-emerge pass (up and back on the same day) and a single post-emerge pass with an M&W Gear minimum-tillage rotary hoe designed to handle heavy residue. But he's hoping that newly adopted cover-crop practices will eliminate the double pre-emerge hoeing.

The Thompsons use a standard Buffalo ridge-till cultivator. They set disk hillers just 5 inches apart at first cultivation, and use 24-inch sweeps. Culti-Vision mirrors from Klingler Manufacturing (Rt. 1 Box 186A, New Ulm MN 56073, (507) 354-8735) help keep them on track, and covered tent shields set close to the ground keep weed seeds from being thrown back into the row. At last cultivation, they use 14-inch sweeps with ridging wings, and set disk hillers 10 inches apart in valley centers, turned in to throw soil on the plants.

Their five-year rotation of corn-soybeans-corn-oats-hay also helps control weeds. In '88, the soybean ground in that rotation contained an average of just one velvetleaf plant for every 2 acres. In cash-grain demonstration plots where the Thompsons grow ridge-tilled corn and beans on a two-year rotation, there were 807 velvetleaf plants per acre.

Covers Check Weeds

The Thompsons began experimenting with cover crops in the early '80s, by overseeding alfalfa and clovers into corn and beans at last cultivation. They soon discovered that perennial legumes were too tough to kill with the planter and competed with crops the next season.

Hairy vetch, a winter-annual legume, showed more promise. "The planter does a good job taking it out," says Dick. "Vetch worked for us a couple of years, but then got shaded out when seeded at last cultivation. We realized that if we were working to shade out weeds, we were also going to shade out the overseeded cover crops."

So in '84, the Thompsons began aerially seeding cover crops into corn and beans in early September. They tried hairy vetch, oats and rye alone and in combinations, and finally settled on a mix containing 30 pounds of vetch, 20 pounds of oats and 15 pounds of rye.

"The vetch needs a companion," says Dick. Oats and rye grow quickly in fall for ground cover, then oats winterkill. The dead mulch protects both soil and vetch. It also provides some allelopathic (natural growth-suppressing) weed control in spring. Total seed and planting cost for the mix was about $21 per acre. "These costs could be reduced by growing your own seed," says Dick.

In spring, vetch and rye regrowth competes with early weeds. Vetch also provides nitrogen, while rye residue suppresses weeds allelopathically. Together, the two crops improve soil structure and increase earthworm numbers, say the Thompsons. But don't be tempted to seed rye more heavily. The rank regrowth will be difficult to kill, use too much water and tie up too much soil N, warns Dick.

In '88, the Thompsons replaced aerial seeding (which cost about $5 per acre) with a highboy broadcast seeder built for $2,500 from a used corn detasseler. It seeded 16 rows per pass.

They still weren't satisfied that broadcasting was the most cost-effective way to establish covers. "The seed cost is too high," says Dick. When broadcast, most of the seed rolls off the ridge and germinates in the valleys between the rows, he explains. There, the cover can't improve soil structure in the ridge, or suppress in-row weeds. It's also hard to kill in spring with the planter. "We now think the key is to drill covers after harvesting corn and beans."

Here's what the Thompsons suggest: After soybean harvest, drill a mix of 10 pounds of hairy vetch and 10 pounds of rye per acre in two rows six inches apart straddling the row atop the ridge. Following corn harvest, shred the stalks and drill 15 to 20 pounds of rye per acre the same way. Shred or graze covers if necessary just before planting in the spring.

Drilling covers on the ridgetop concentrates their soil-enhancing and weed-suppressing properties right where they're needed most, says Dick. "And when you drill them, you can get them into moist soil. You can use lower seeding rates and they'll come up fast."

Best of all, his cost of establishing rye this way is just $8.30 per acre (including $6.50 for drilling), Dick estimates. He uses a Great Plains drill, but almost any drill will do as long as wheels and drill units match ridge spacing, he says.

In '91 replicated trials, the Thompsons learned that rye drilled on the ridges still helped control weeds between the rows when it was pushed into the centers by the planter. "Root exudates from rye drilled in the ridgetops may also eliminate the need for pre-emerge rotary hoeing, allowing you to postpone the decision to hoe until after the crop is up," says Dick.

150-Bushel Oats

The Thompsons' basic rotation hasn't changed much over the years. "We just picked up Dad's old five-year rotation in '67 and that was it," recalls Dick. They're still refining the details, though. For example, in hayfields to be planted to corn, they usually disk twice in fall, then build ridges and drill a cover crop. In '88, they left some hay ground undisturbed until spring, when they incorporated manure and planted corn into a flat seedbed. That method yielded about 3.5 bushels more per acre compared with fall-built ridges. But Dick considers spring plowing too risky.

Reducing drought risk is a major goal for the Thompsons. Although they weren't hit by the '88 drought as badly as some, they say their soil-building rotation and cover crops

Drilling cover crops in two rows six inches apart straddling the ridgetop reduces seeding rates, makes it easier to kill covers with the planter and concentrates their weed-killing and soil-improving benefits where they're needed most.

helped them weather that season better than most other Midwest farmers. From '85 to '87, their soybeans averaged 5.4 bushels per acre more than the county average, and their corn averaged 8.7 bushels less. In '88, beans were 17 bushels above county averages, and corn was 27 bushels more. The farm received 0.8 inch of rainfall during a 15-minute period on May 9. "It went in the ground here, while all around us it ran off and ponded," notes Sharon.

Better infiltration of moisture may have been enough to offset the water-robbing effect of the Thompsons' cover crops that year in most fields. But in one cornfield, yields were down 20 bushels per acre where the planter left some rye close to the corn row. "I blame most of that on the rye that didn't get taken out with the planter, not on the cover itself," says Dick. He believes drilling only two rows of rye on soybean ridges will eliminate this problem.

Thanks to early moisture and a few "mistakes" in '88, the Thompsons learned they could nearly double their usual oats yield. First, the co-op broadcast 120 pounds of potash per acre — double what the Thompsons ordered. Then Dick accidentally drilled 4.25 bushels of oats instead of the usual 3 bushels. The short-strawed DON oats averaged 142.5 bushels per acre, and tested at 38 pounds per bushel. "Some areas of the field were well over 150 bushels," says Dick. "I didn't think it was possible to grow that many oats and still have the hay seeding take, too."

Half-Price Hog Housing

The Thompsons' 70-sow, farrow-to-finish operation and 80-cow, cow-calf herd are as essential to their farming system as small grains, cultivators and cover crops. "Livestock eat the hay and oats and their manure fertilizes our crops," says Sharon. "Some people think it's hard to justify pasturing cows on high-priced corn ground. But it pencils out well when you consider the whole system."

Livestock also deserve some credit for motivating the Thompsons to improve their farming methods. The couple used to confine cattle to feedlots, farrow hogs in a remodeled granary and grow continuous corn. That era began in '58, when Dick returned to the home place armed with a master's degree in animal science from Iowa State University (ISU). And it didn't end until the Thompsons quit chemicals in '67. "We had the 'expand-and-get-specialized disease' as bad as anyone," recalls Dick.

Disease was the rule with their livestock, too, he continues. "We did everything we could, including putting antibiotics in the feed and water. All we did was breed resistant diseases. If one pig got sick, they all got sick. And more drugs didn't help. Resistant diseases were showing up in the cattle."

The cows are healthier now that they calve on 40 acres of pasture that's part of a six-year rotation of corn-soybeans-oats-meadow. Hogs are raised in open-air farrowing, nursery and finishing units designed by Cargill during the '60s. The Thompsons built the facilities during the summers of '78 and '79, and added two nursery pens in '84, for a total cost of about $63,000. That's less than half what they would have spent to build total-confinement facilities for the same number of hogs.

The Thompsons use modified Cargill housing as part of an integrated hog production system that eliminates the need for subtherapeutic antibiotics.

The Thompsons' innovative hog-management methods are equally important to herd health. They feed ground ear corn, oats and iron, and limit starter rations to just 16.5 percent crude protein instead of the more common 21 percent. Higher-protein starter rations (fed largely to keep lysine levels high) seem to promote scours in young pigs, according to the Thompsons' research. To be sure their pigs get enough lysine, the couple supplement starter rations with this vital amino acid at the rate of 2.5 pounds per ton.

In place of routine subtherapeutic antibiotics, the Thompsons feed probiotics and other materials containing live, beneficial bacteria found naturally in an animal's gut. These are added to rations at various rates, to help the animals resist disease and to improve sows' milking ability. "When you use antibiotics, things get out of kilter in the hog's gut," says Dick. "We use probiotics to maintain a normal balance of flora in the animal's digestive tract."

Cattle eat home-grown oats and hay, graze alfalfa and provide manure. "Some people think it's hard to justify pasturing cows on high-priced corn ground. But it pencils out well when you consider the whole system," says Sharon.

Still, he adds, probiotics alone are no cure-all for disease and feeding problems. "They have to be part of a total hog-management system." The couple also spread ag lime on pen floors to raise pH and kill pathogenic organisms. To help gestating sows develop immunity to diseases, they spread a loader bucket of manure from the farrowing unit in the gestation pens every week.

Weaning rates are up to about 8.5 pigs per litter. "Diseases haven't disappeared, but they've stabilized," says Dick. "With this system, we're able to handle crises."

The Thompsons' beef herd is a three-way composite cross of a quarter Red Holstein (for milking ability), a quarter Hereford (for durability) and half Red Angus (for meatiness). The herd was shifted between two 20-acre pastures until '89, when the Thompsons started subdividing each pasture into three paddocks. The increase in forage production was dramatic. "We should probably subdivide more and move cows more, but we have hay coming out of our ears," says Dick. "We need more cows to use up all the forage, so I'll start holding back more heifers."

Bunker Saves K

Dick and Sharon are quick to share their failures as well as their successes, so other farmers can build on what the Thompsons have learned. For example, while manure probably saved the farm when they quit chemicals cold-turkey in '67, they've only recently learned to perfect its storage and management.

"We were fortunate our yields didn't fall out of bed when we cut out the fertilizer," recalls Sharon. "Luckily, we still had manure in the system." The problem was, they treated manure like a waste product, spreading it in winter on top of the snow. "It was bad news. I'm sure we lost most of the nitrogen. All I can say is we got rid of the manure," says Dick.

After a few years, corn yields following soybeans started slipping from 120 bushels per acre to 80 to 90 bushels. About that time, the Thompsons decided to start composting manure. They unloaded it into windrows and turned it three times by reloading and unloading it with a loader and manure spreader. The tedious process took 30 hours of labor to make 300 tons of compost, but it did the trick. "Yields jumped back up. It made the compost look great," says Dick, noting that fall-spread compost probably supplied more N than winter-spread manure did.

But after eight years of composting, crops started showing signs of potassium deficiency. "We couldn't get enough compost on the fields to keep the yields up," Dick recalls. "It turns out some of our potassium was leaching out of the windrows and never getting back into the field."

To save potassium-rich liquids, the Thompsons built a 3,755-cubic-yard concrete bunker in '86. It holds a full year's worth of manure from hog and cattle operations, plus an equal amount of sewage sludge from nearby Boone, a city of 12,000. Cement kiln dust, which is added at the waste treatment plant, stabilizes the sludge and provides additional K. The Thompsons regularly test sludge before application to avoid possible contamination by heavy metals.

The Thompsons are almost as famous for their on-farm research as they are for their innovative farming practices. Early on, they used simple test strips to compare practices. But they soon found out their experiments had no credibility with university researchers and, worse, could be unreliable and misleading.

Dick Thompson — who has shared his vision of regenerative agriculture with secretaries of agriculture, Congress and countless university scientists, USDA officials and international agriculturalists — prepares his corn planter for another season.

Research For Farmers

So, in cooperation with Rodale Institute researchers and other scientists, they developed a design that satisfies researchers' needs for statistical validity, yet is still practical for farmers. The design compares two farming practices — say, manure vs. purchased fertilizer — in pairs of field-length strips. Each pair is repeated six to eight times. Placement of the compared practices in each pair is chosen at random.

On-farm research cooperators with Practical Farmers of Iowa have proved the technique to be even more reliable than university-style small-plot experiments. "And ours costs less too," notes Dick.

Now, researchers come to the Thompsons for the opportunity to study their practices. One of the most interesting ongoing studies is being conducted by USDA's Soil Tilth Lab, which is comparing the Thompsons' fields with neighboring conventionally farmed fields for differences in soil health and the fate of nitrates and pesticides.

From The Heart

The Thompsons wonder if they'd still be farming today had they not quit relying on chemicals in '67. "We had a lot of borrowed money, then. The bank wasn't worried. But we knew what had to be paid back," says Dick. "It wasn't a crisis, but it would have been if we'd kept farming the way we did."

Today, the Thompsons are debt-free. But the changes they made affected more than their balance sheet. "There's a spiritual side to farming, too," says Sharon. "We just weren't satisfied with the way we felt inside about how we were farming, then."

They hope farmers catch that part of their message, too. "When the economics work with regenerative farming, people will go along with it," says Sharon. "But what happens when the economics are different? How long-lasting will changes be if they're only motivated by money? There also has to be a commitment from the heart."

Appendix: Other Information Sources

Before cutting inputs on many acres, you might want to check with experienced sustainable farmers in your area. A good place to start is with the nearest Rodale Institute network cooperator. The Institute has networks in the Midwest, Mid-Atlantic and Mid-South regions. For information about our workshops and field days (including those held at the Rodale Institute Research Center), write: Field Days, Rodale Institute, 222 Main St., Emmaus PA 18098.

Many other organizations also sponsor field days and workshops or provide information about farming with fewer chemicals. Check the following list for those near you.

National and International

Alternative Farming Systems Information Center
National Agricultural Library
Room 109-C
10301 Baltimore Blvd.
Beltsville MD 20705
(301) 504-6559

American Farmland Trust
1920 N Street N.W.
Washington DC 20036
(202) 659-5170

Appropriate Technology Transfer for Rural Areas (ATTRA)
P.O. Box 3657
Fayetteville AR 72702
(800) 346-9140

Bio-Dynamic Farming and Gardening Association
P.O. Box 550
Kimberton PA 19442
(215) 935-7797

Bio-Integral Resource Center
P.O. Box 7414
Berkeley CA 94707
(510) 524-2567

Conservation Technology Information Center
1220 Potter Drive, Room 170
West Lafayette IN 47906-1383
(317) 494-9555

Institute for Alternative Agriculture
9200 Edmonston Road, Suite 117
Greenbelt MD 20770
(301) 441-8777

International Alliance for Sustainable Agriculture
1701 University Ave. S.E.
Minneapolis MN 55414-2076
(612) 331-1099

National Coalition Against the Misuse of Pesticides
701 E St. S.E., Suite 200
Washington DC 20003
(202) 543-5450

National Organic Standards Board
Dr. Hal Ricker
USDA/AMS/TMD
Room 4006-S
P.O. Box 96456
Washington DC 20090-6456
(202) 720-2704

Organic Crop Improvement Association
3185 Twp. Rd. 179
Bellefontaine OH 43311
(513) 592-4983

Organic Farmers' Associations Council
Route 3 Box 494
Stage Coach Road
Siler City NC 27344
(919) 472-4672

Organic Foods Production Association of North America
P.O. Box 1078
Belchertown MA 01301
(413) 774-7511

Rodale Institute
222 Main St.
Emmaus PA 18098
(215) 967-8405

Sustainable Agriculture Research and Education
USDA/CSRS
342 Aerospace Bldg.
Washington DC 20250-2200
(202) 401-4640

USDA Office for Small-Scale Agriculture
328 Aerospace Bldg.
Washington DC 20250-2200
(202) 401-1805

Winrock International
Institute for Agricultural Development
Route 3 Box 376
Morrilton AR 72111
(501) 727-5435

World Sustainable Agriculture Association
MOA Foundation — Los Angeles
8564 Melrose Ave.
West Hollywood CA 90069
(213) 657-7200

Midwest

Alternative Energy Resources Organization
25 S. Ewing St., Suite 214
Helena MT 59601
(406) 443-7272

Center for Rural Affairs
P.O. Box 406
Walthill NE 68067
(402) 846-5428
and:
P.O. Box 736
Hartington NE 68739
(402) 254-6893

Illinois Sustainable Agriculture Society
Box 418
Ashland IL 62612
(217) 476-3332

Indiana Sustainable Agriculture Association
Krannert Building Room 649
Purdue University
West Lafayette IN 47907-1145
(317) 494-4245

Iowa Organic Growers & Buyers Association
2517 Park Ave.
Muscatine IA 52761
(319) 263-5701

Kansas Rural Center
P.O. Box 133
Whiting KS 66552
(913) 873-3431

The Land Institute
2440 E. Water Well Road
Salina KS 67401
(913) 823-5376

Land Stewardship Project
P.O. Box 130
Lewiston MN 55952
(507) 523-3366
and:
103 West Nichols Ave.
Montevideo MN 56265
(612) 269-2105

Leopold Center for Sustainable Agriculture
126 Soil Tilth Bldg.
Iowa State University
Ames IA 50011
(515) 294-3711

Michael Fields Agricultural Institute
3293 Main St.
East Troy WI 53120
(414) 642-3303

Michigan Agricultural Stewardship Association
c/o Jerry Wirbel
4240 Stark Road
Hope MI 49628
(517) 689-3857

Minnesota Department of Agriculture
Energy and Sustainable Agriculture Project
90 W. Plato Blvd.
St. Paul MN 55107
(612) 297-1320

Minnesota Institute for Sustainable Agriculture
411 Borlaug Hall
1991 Buford Circle
St. Paul MN 55108-1013
(612) 625-8235

Organic Growers and Buyers Association
1405 Silver Lake Road
New Brighton MN 55112
(612) 636-7933

Missouri Organic Association
19010 S. Mackie Lane
Hartsburg MO 65039
(314) 657-4156

Missouri Sustainable Agriculture Society
c/o Farm Alliance of Rural Missouri
P.O. Box 1094
Jefferson City MO 65102
(314) 636-6005

Nebraska Sustainable Agriculture Society
P.O. Box 736
Hartington NE 68739
(402) 254-2289

Northern Plains Sustainable Agriculture Society
Susanne Retka Schill
Box 36
Maida ND 58255
(701) 256-2424

Ohio Ecological Food and Farm Association
65 Plymouth St.
Plymouth OH 44865
(419) 687-7665

Practical Farmers of Iowa
Rick Exner or Gary Huber
2104 Agronomy Hall
Iowa State University
Ames IA 50011
(515) 294-1923

South Dakota Sustainable Agriculture Society
R.R. 2 Box 37
Rosholt SD 57260
(605) 537-5971

Sustainable Agriculture Center
Chuck Francis
University of Nebraska
219 Keim Hall
Lincoln NE 68583-0910
(402) 472-2056

Sustainable Farming Association of Minnesota
P.O.Box 367
Montevideo MN 56440
(612) 269-2105

Wisconsin Rural Development Center
1406 Highway 18-151 East
Mount Horeb WI 53572
(608) 437-5971

Wisconsin Sustainable Agriculture Program
Department of Agriculture, Trade & Consumer Protection
P.O. Box 8911
Madison WI 53708
(608) 273-6210

Northeast

Farming Alternatives Project
Northeast Organic and Sustainable Farmer Network
Warren Hall
Cornell University
Ithaca NY 14853
(607) 255-9832

Finger Lakes Organic Growers Cooperative Inc.
P.O. Box 549
Trumansburg NY 14886
(607) 387-3333

Maine Organic Farmers & Gardeners Association
P.O. Box 2176
Augusta ME 04338
(207) 622-3118

Maryland Organic Food and Farming Association
6201 Harley Road
Middletown MD 21769-6516
(301) 371-4814

Mountain State Organic Growers and Buyers Association
P.O. Box 642
Morgantown WV 26507
(304) 249-5139

Northeast Organic Farming Association of Connecticut
P.O. Box 386
Northford CT 06472
(203) 484-2445

Northeast Organic Farming Association of Massachusetts
Julie Rawson or Jack Kittredge
411 Sheldon Rd.
Barre MA 01005
(508) 355-2853

Northeast Organic Farming Association of New Hampshire
c/o White Farm
150 Clinton St.
Concord NH 03301
(603) 648-2521

Northeast Organic Farming Association of New Jersey
31 Titus Mill Road
Pennington NJ 08534
(609) 737-6848

Northeast Organic Farming Association of New York
P.O. Box 21
South Butler NY 13154
(315) 365-2299

Northeast Organic Farming Association of Rhode Island
89 Country Drive
Charlestown RI 02813
(401) 364-9930

Northeast Organic Farming Association of Vermont
15 Barre St.
Montpelier VT 05602
(802) 229-4940

New England Small Farm Institute
P.O. Box 937
Belchertown MA 01007-0937
(413) 323-4531

Pennsylvania Association for Sustainable Agriculture
P.O. Box 316
118 West Main St.
Millheim PA 16854
(814) 349-9856

Working Land Fund
P.O. Box 190
South Strafford VT 05070
(802) 765-4506 or 4662

South

Arkansas/Oklahoma Agriculture Network
Arkansas Cooperative Extension Service
P.O. Box 391
Little Rock AR 72203
(501) 671-2000

Carolina Farm Stewardship Association
115 W. Main St.
Carrboro NC 27510
(919) 968-1030

Community Farm Alliance of Kentucky
200 Short St. #10
Berea KY 40403
(606) 986-7400

Florida Organic Growers Association
Route 1 Box 116
Jennings FL 32053
(904) 938-2045

Georgia Organic Growers Association
P.O. Box 567661
Atlanta GA 31156
(404) 621-4642

Kentucky Organic Farmers Association
1735 Sulphur Lick Road
New Hope KY 40052
(502) 692-2339

Kerr Center for Sustainable Agriculture
P.O. Box 588
Poteau OK 74953
(918) 647-9123

Mississippi Organic Growers Association
Tom Dana
277 Hurricane Creek
Lumberton MS 39455
(601) 796-4406

Ozark Organic Growers Association
P.O. Box 1211
Fayetteville AR 72702
(501) 521-0206

Ozark Small Farm Viability Project
P.O. Box 99
Mt. Judea AR 72655
(501) 434-5594

South Carolina Organic Association
129 Organic Lane
West Columbia SC 29172
(803) 791-5733

Tennessee Alternative Growers Association
Route 1 Box 77
Bloomington Springs TN 38545
(615) 653-4402

Tennessee Land Stewardship Association
Jim Joiner
Route 1
Liberty TN 37095
(615) 563-2353

Texas Department of Agriculture
Agriculture Resources Program
P.O. Box 12847
Austin TX 78711
(512) 463-7612

Texas Organic Growers Association
Route 1 Box 9
Hutto TX 78634
(512) 759-3741

Virginia Association of Biological Farmers
c/o John Burns
P.O. Box 252
Flint Hill VA 22627
(703) 675-3982

West

California Action Network
P.O. Box 464
Davis CA 95617
(916) 756-8518

California Association of Family Farmers
P.O. Box 363
Davis CA 95617
(916) 756-7420

California Certified Organic Farmers
P.O. Box 8136
Santa Cruz CA 95061
(408) 423-2263

Center for Holistic Resource Management
P.O. Box 7128
Albuquerque NM 87194
(800) 654-3619

Committee for Sustainable Agriculture
P.O. Box 1300
Colfax CA 95713
(916) 346-2777

Idaho Organic Producers Association
11741 Bullock Land
Middleton ID 83644
(208) 585-6140

New Mexico Organic Growers Association
1312 Lobo Place N.E.
Albuquerque NM 87106
(505) 268-5504

Northwest Coalition for Alternatives to Pesticides
P.O. Box 1393
Eugene OR 97440
(503) 344-5044

Oregon Tilth
P.O. Box 218
Tualatin OR 97062
(503) 691-2514

Progressive Farmers of the Inland Northwest
Bob Klicker
R.D. 4 Box 236
Walla Walla WA 99362
(509) 525-2494

Sustainable Agriculture Research and Education Program
c/o Agronomy and Range Science Department
University of California
Davis CA 95616
(916) 752-7556

Washington Tilth
c/o Anne Schwartz
1219 East Sauk Road
Concrete WA 98237
(206) 853-8449

Canada

Canadian Organic Growers Inc.
Box 6408 Station "J"
Ottawa, Ontario K2A 3Y6
(613) 256-1848

Centre de Developpement D'Agrobiologie du Quebec
224 Rue Principale
Ste. Elizabeth-de-Warwick, Quebec J0A 1M0
(819) 358-3850

Ecological Agriculture Projects
Box 191
Macdonald College of McGill University
Ste. Anne de Bellevue, Quebec H9X 3V9
(514) 398-7771

Resource Efficient Agricultural Production (REAP Canada)
Roger Samson
P.O. Box 125
21111 Lakeshore Road
Ste. Anne de Bellevue, Quebec H9X 1C0
(514) 398-7743

Pesticide Action Group Canada
P.O. Box 22021
Westmount Postal Outlet
Waterloo, Ontario N2L 6J7